Urban Sustainability

The Urban Sustainability Book Series is a valuable resource for sustainability and urban-related education and research. It offers an inter-disciplinary platform covering all four areas of practice, policy, education, research, and their nexus. The publications in this series are related to critical areas of sustainability, urban studies, planning, and urban geography.

This book series aims to put together cutting-edge research findings linked to the overarching field of urban sustainability. The scope and nature of the topic are broad and interdisciplinary and bring together various associated disciplines from sustainable development, environmental sciences, urbanism, etc. With many advanced research findings in the field, there is a need to put together various discussions and contributions on specific sustainability fields, covering a good range of topics on sustainable development, sustainable urbanism, and urban sustainability. Despite the broad range of issues, we note the importance of practical and policy-oriented directions, extending the literature and directions and pathways towards achieving urban sustainability.

The series will appeal to urbanists, geographers, planners, engineers, architects, governmental authorities, policymakers, researchers of all levels, and to all of those interested in a wide-ranging overview of urban sustainability and its associated fields. The series includes monographs and edited volumes, covering a range of topics under the urban sustainability topic, which can also be used for teaching materials.

Ali Cheshmehzangi

Thailand's Tropical Thresholds

Coastal Climate Resilience and Sustainable Urban Transformations

Springer

Ali Cheshmehzangi
Architecture, Design & Planning
The University of Queensland
Brisbane, QLD, Australia

ISSN 2731-6483 ISSN 2731-6491 (electronic)
Urban Sustainability
ISBN 978-981-95-4974-0 ISBN 978-981-95-4975-7 (eBook)
https://doi.org/10.1007/978-981-95-4975-7

This Springer imprint is published by the registered company Springer Nature Singapore Pte Ltd.
The registered company address is: 152 Beach Road, #21-01/04 Gateway East, Singapore 189721, Singapore

In between mangroves of Rai Leh

Land and life in laid-back Rawai

With multi-colours of Karon

From Phetchaburi to Cha-Am

Coastal areas of Hua Hin

Edges and thresholds of Krabi

The slow arrival of storm, Rai Leh (Railay) Island, Krabi

Crossing the western coastal areas of Thailand

Off the Mucho Amor coastal area, South-Western Phuket Island

Not your everyday scene in Bang Tao coastal area

The extended blue and green of Surat Thani

And there is more on the table in Krabi.

Photos taken by Ali Cheshmehzangi, Thailand, Aug 2025

It is safe to say that like any global academic with a strong emphasis on equity, I stand stronger on the shoulders of, and beside, many remarkable women. Throughout my academic journey, I have guided many emerging female scholars, practitioners, and colleagues. The impact of these relationships is mutual, i.e., my mentorship helps develop strong women, and their engagement, vision, and resilience no doubt strengthen my work and leadership in turn. To many strong women whom I got to know so far as family members, friends, colleagues, and collaborators, all in various forms of blessings on my path: Afsaneh, Juan, Tree, Ayra, Sara, Farnoush, Fereshteh, Zohreh, Zhaohui, Sue, May, Lydia, Tong, Anna, Susan, Olivia, and many more.
This book is particularly dedicated to Mrs Olivia Dwyer, my School Manager at the School of Architecture, Design and Planning (ADP), The University of Queensland; for her commitment and unwavering support that have been instrumental throughout these very busy, yet, progressive and productive, few months. Your dedication behind the scenes

has not only kept the wheels turning smoothly but has provided me with additional backing for new ideas to flourish and progress to take shape. On behalf of the School, I want to make sure I express my deep gratitude and respect to you as I reflect on our shared pursuit of excellence and advancement.

Preface

"Balancing the beauty of Thailand's coasts with the needs of its people".

I just want to start this rather sentimental and metaphysical brief dialogue here by reminding us that '*just because we cannot do it, it doesn't mean it cannot be done*'. I reminded myself of this agitating thought with many possibilities and opportunities during my site visits, extended fieldwork, and engagement sessions and workshops in various parts of Thailand.

Now, let's begin! In some parts of the planet, nature and water speak in quiet, archaic voices. These discussions have long influenced the rhythms of daily life throughout Thailand's southern and eastern coasts, with fisherman throwing nets before dawn, mangrove roots embracing silt and salt, and tidewaters receding into tales passed down from grandmother to kid. However, these classic rhythms have been disrupted. Not only is the physical landscape shifting, but so are the political, social, ecological, and emotional spheres in which we all inhabit.

This book was born out of a profound concern, but also out of hope. Thailand, a nation so intricately linked to its coastal ecosystems, is currently at a turning point in its history, a tropical threshold where choices made now will influence future generations. I have travelled along these coastlines and heard the opinions of scientists, planners, farmers, fishermen, young people, knowledge holders, local guides, and seniors. I've encountered questions that don't have straightforward answers. How can we preserve these beaches' natural beauty while preserving the life they support? When the ground beneath us, physically and figuratively, is changing, how can we plan, direct, and adjust?

The title *Thailand's Tropical Thresholds* is not a metaphor chosen lightly. A threshold denotes both a point of entry and a decision-making moment. It's the intersection between the past and the future. The mangrove line dividing the sea from the land, the industrial corridors of Chonburi and Rayong, where urban development encroaches upon delicate ecosystems, the voices of local communities resisting, adapting, and innovating, the silent resilience of coral reefs and seagrass beds, and the eyes of children learning to read the tides anew in an uncertain world are all examples of thresholds that appear along Thailand's coastlines.

Climate change has rendered these thresholds more precarious. Rising seas tell tales of flooded homes, lost livelihoods, and drowned memories; they are more than just an ecological statistic. The effects of warming temperatures are tangible; farmers' hands are cracked by salt, and coral bleaching events cause death where color formerly was. Despite this vulnerability, Thailand's coastlines present a powerful invitation to envision a future that marks a shift toward regeneration rather than just a continuation of the current catastrophe.

This book aims to provide a thoughtful, modest, and lucid response to that call. No book can claim to have all the answers, so it does not. However, it collects lessons, narratives, tactics, and insights from Thailand's most climate-vulnerable areas, including the estuarine ecosystems of Phetchaburi, where livelihoods and biodiversity are closely intertwined; the policy laboratories of the Eastern Economic Corridor (EEC), where tensions between industrial growth and local identity emerge; and the village of Ban Nai Nang in Krabi, where intergenerational adaptation flourishes.

I have always felt that people, people who recall, resist, and reimagine, are more important to the future of cities and coasts than technology or planning theory. I am a global scholar, and hopefully a leading one, currently at a leading global university and previously in other locations. Although my academic journey has brought me across continents, I have seen a unique fusion of modern resilience thinking and ancient ecological understanding in Southeast Asia, especially in Thailand. This convergence is delicate. It must be nurtured, not commodified; respected, not romanticized. I have worked with many people of various positions and levels in many difficult locations, and I continue to do so, as I believe I carry a rather heavy load of social responsibility, a sort of social agency (if I may) that requires our work to be meaningful and impactful, and not of the nature of accumulated citations and multiple research grants and papers. To me, this remains central in conducting research in places that need our presence, contributions, and dedication.

In this book, I explore multiple dimensions of climate resilience. I delve into governance architectures, which are frequently disjointed, occasionally enabling, and always significant. We consider nature-based solutions as lived, location-specific adaptations rather than as fad interventions. We look at how decentralized governance, gender-inclusive planning, and youth-led digital technologies may give adaptation initiatives credibility and durability. We elevate the silent authority of communal knowledge and challenge the top-down imposition of national development agendas.

Thailand's future will be formed by its imagination as much as its infrastructure. The capacity to blur the lines between the planners and the people, between the formal and the informal, and between what is required and what is feasible will be crucial. Building stronger seawalls is only one aspect of a climate-resilient Thailand; another is restoring trust between communities and institutions, between generations, and between people and nature.

Urbanization will continue to rise in the ensuing decades. Cities along the coast will expand. There will be increasing pressure on people, land, and water. But the opportunities will, too. Opportunities to create coastal cities that don't fight nature but rather coexist with it. Possibilities to enable local governments to co-author locally anchored futures rather than just carrying out national plans. Possibilities to rebuild damaged landscapes as investments in survival and dignity rather than as artistic endeavors.

One of the most powerful lessons I have learned through the writing of this book is this: Resilience is not a fixed endpoint. It is a process, a practice of learning, of listening, of adapting. It emerges not only in grand policy frameworks but in the smallest of actions: a child planting a mangrove, a grandmother sharing weather patterns, a village redrawing its map with a future in mind.

This book is written for those who stand at the frontlines of change, planners, policymakers, architects, researchers, educators, students, and above all, communities. It is intended to be provocative as well as a reference. A resource for anyone wishing to comprehend the uniqueness and intricacy of Thailand's coastal resilience environment; a challenge for those prepared to challenge conventions and jointly forge new directions.

I dedicate this work to everyone who cares for Thailand's coasts, not just in terms of profession or policy, but also in terms of spirit. To those who fish at the crack of dawn, manually plant mangroves, speak up at village meetings, carefully craft zoning laws, and digitize knowledge for future generations. The tide underlying this text is your stories.

As we move forward, we must carry two truths. First, that the climate crisis is real, urgent, and unrelenting. And second, that within every threshold lies the possibility of transformation, only if we are willing to meet it with wisdom, creativity, and resolve.

Let this book be one such meeting place.

April 2025

Professor Ali Cheshmehzangi
Head of School of Architecture, Design and Planning
The University of Queensland
Brisbane, QLD, Australia

Acknowledgement As always, I would like to acknowledge those who made another project a reality. Our earlier (general) discussions on this book project in April 2023 and then in late November 2024 in Chiangmai and further detailed thoughts about making a journey to Thailand in August 2025 made everything possible. It seems like yesterday when I first arrived in Thailand in July 2012, but over many years of engagement with multiple stakeholders through several research projects, I have gained more knowledge about the country and more respect for its wonderful people; and now have a better understanding of the context, better than I initially had in mind.

All those studies helped me a lot to have a broader perspective of sustainability and sustainable transformations of particular regions in this magnificent country. There remain some critical views of course, but in return, there are so many purpose-driven solutions we have provided over the years. From 'cooling design' perspectives to 'microclimate' strategies, 'urban morphology' and 'urban form' analytical research as well as the extended 'energy and tropical housing' work we did between 2013 to 2017 with Dr. Terry Thomas of Warwick University under the ELITH project, 'rural development' evaluation in several regions, 'vernacular architecture' and energy solutions we worked on during those wonderful years, 'low-carbon transitions' in cities and regions, 'density' and 'urban green infrastructures' with my colleague, Ar. Chris Butters, and now urban and costal transformations as part of 'tropical thresholds'; all have been excellent and memorable experiences. Thailand truly offers a lot for us to learn, study, and contribute to, something that we should remember as part of our social agency or social responsibility in giving back to societies and communities around the globe – with no exceptions, with no hesitation!

I also like to acknowledge those who helped with several successful and eventful visits in 2023, 2024, and 2025; those who participated in our engagement sessions, those who helped with some of our data collection, and those who allowed us to access sites, visit projects, and engage with various groups of people. In all my experiences, people of Thailand stand out as being extremely welcoming, supportive, and encouraging. These were part of capturing the realities on the ground in all our visits and fieldworks over these wonderful short journeys to many Thai cities, townships or towns, coastal regions, and rural areas. With all my heart, I give my respect to all places we have been to in Thailand and their diverse people, landscapes, nature, and communities.

Competing Interests The author has no competing interests to declare that are relevant to the content of this manuscript.

About This Book

This book dives into in-depth case studies across Thailand's low-lying coastal regions and its rapidly changing urban centers. It focuses on main coastal provinces and regions, where multidisciplinary research shows how hydro-meteorological hazards (such as storm surges, flash floods, erosion) are reshaping both natural environments and human systems. Detailed physico-geomorphological modelling reveals dramatic rates of erosion and accretion, up to -34.5 m/year on mangrove coastlines, versus -4 m/year on sandy beaches, with projections of increased storm severity and flooding because of sea-level rise.

Comparing local adaptation capacity, socioeconomic analysis reveal that some populations exhibit more robust social networks, whereas urbanized regions exhibit higher infrastructure resilience. Both show comparable total coastal vulnerability indices despite these variations. Through the integration of livelihood surveys, public health assessments, and community governance audits, the book reveals grass-roots coping strategies, ranging from cooperative mangrove restoration to unofficial evacuation plans.

On the urban front, the book explores Bangkok and other coastal cities grappling with compound threats: land subsidence, urban flooding, and heat stress. It assesses infrastructure, such as drainage systems, green-blue networks, smart early warning systems, and policy frameworks governing land-use planning and disaster response.

The book provides a comprehensive perspective on resilience changes by fusing case-study narratives with institutional analysis, policy evaluation, and UN/Green Climate Fund-backed initiatives. In addition to concrete infrastructure, it highlights nature-based solutions (such as mangrove replanting, sand fences, and crab banks), highlighting the connections between sustainable urban development, multi-scale governance, and local knowledge. In sum, *Thailand's Tropical Thresholds* offers a detailed yet comprehensive picture of urban and coastal resilience, making it an essential tool for academics, decision-makers, and practitioners involved in climate adaptation in tropical coastal regions.

This book with all its case study chapters:

- Showcases rich case studies from Thailand's coastal cities bridging science, policy, and local knowledge.
- Integrates nature-based solutions with urban resilience strategies tailored to tropical coastal conditions.
- Connects grassroots adaptation with national climate policies through actionable, place-specific insights.

The figure provides an overview of the case studies and their geographic distribution across coastal provinces in Thailand's eastern and southern regions.

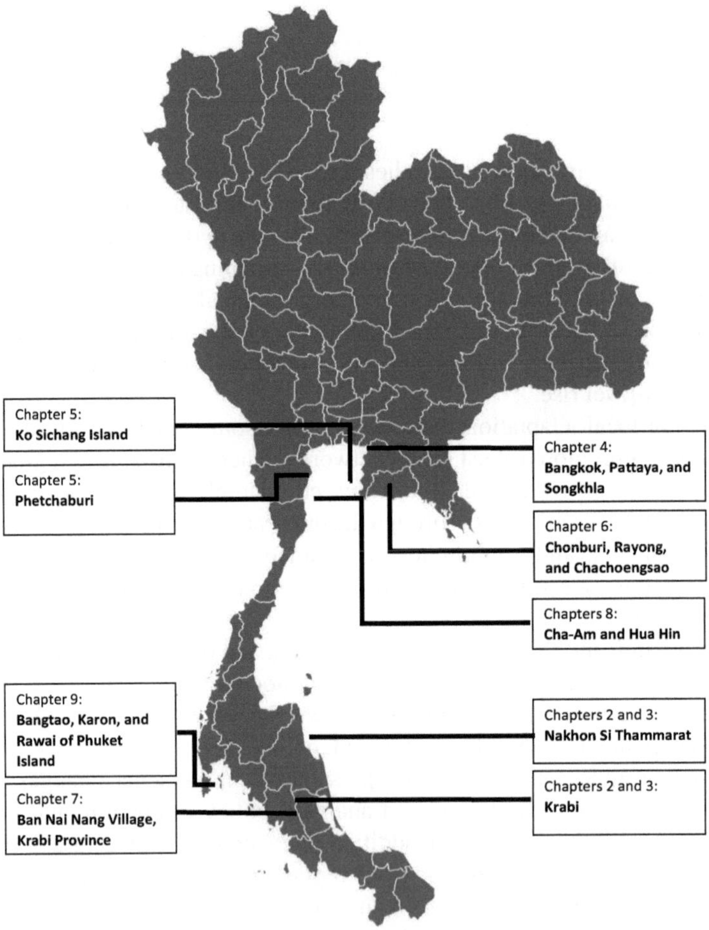

Contents

About the Author

Ali Cheshmehzangi is Professor and Head of the School of Architecture, Design and Planning (ADP) at The University of Queensland. He is also Adjunct Professor at the School of Architecture and Urban Planning at Hunan University. He is the Brisbane Olympics Committee member for Games Legacy and Growth Sub-Committee as well as the Expert member of the 2032 Games led by UQ. He has been in the World's top 2 percent field leader since 2021, recognized by Stanford University. He is among the top 20 global scholars in the urban sustainability research area, as well as the founder and Editor-in-Chief of the Urban Sustainability (US) Book Series with Springer.

Ali has a wealth of experience, expertise, and a strong commitment to innovation and excellence in the fields of architecture, design, and planning, the three core disciplines of his School at UQ. With a career spanning over two decades, he has made significant contributions to the academic and professional communities, with a focus on sustainable and environmentally conscious design. Prior to joining UQ, Ali held several strategic leadership and senior managerial roles, such as Vice-President for International Engagement and Global Partnership, Founding Director and Head of the Center for Innovation in Education and Research, Head of the Department of Architecture and Built Environment, Founding Director of the Urban Innovation Lab, Director of a university-wide Teaching and Learning platform, Director of International Research Network for Rural and Urban Development, Founder and Head of Research Group for Sustainable Built Environment, co-director of university-wide research priority areas, Director of Center for Sustainable Energy Technologies, and Interim Director of Digital Design Lab. So far, Ali has published over 600 journal papers, articles, conference papers, book chapters, and reports. This is his 47th academic book, alongside his many poetry collections that he has written in Parsi/Persian and English. Some of his previous academic books have received awards at the national, provincial, and municipal levels. He also has received international awards and recognition for his research on urban resilience studies and sustainability research, as well as a 2018 Vice-Chancellor's award for his impactful contribution to higher education.

Chapter 1
At the Edge of Change: Framing Thailand's Tropical Thresholds and Climate Resilience

1.1 Locating the Thresholds: Thailand in a Changing Climate

Thailand finds itself at a climatic crossroads. The country is experiencing a complex confluence of demographic, infrastructure, and environmental changes as it straddles the disputed boundary of Southeast Asia's urban-industrial growth and the monsoonal tropics. The southern and eastern regions, which are characterized by water-rich ecosystems, coastal vulnerabilities, and escalating development pressures, are where this is most noticeable (Srinivasan et al., 2024). These are frontlines of possibility rather than merely areas of risk. These areas become vital test sites for how communities and cities adjust, change, and eventually prosper in the face of accelerating climate change.

This book starts with the understanding that Thailand's tropical regions are going through threshold moments—critical junctures where institutional responses, human settlements, and environmental systems are pushed past their shock-absorbing capacity—rather than merely gradual changes. These "tropical thresholds," as this chapter introduces the phrase, denote turning and metamorphic points. In addition to being ecological, they are also social and political. Physical thresholds are indicated by heat waves, sea level rise, saltwater intrusion, and more erratic rainfall patterns. However, there are also boundaries in terms of community resilience, governance, and the cultural perception of what a liveable future looks like.

Thus, the urgency and scope of this book are framed by the idea of tropical thresholds. Tropical climates, particularly those in coastal and monsoon-influenced regions, require a different perspective than temperate or dry ones. Heat is a structural force rather than a seasonal visitor. Water is abundant and erratic, not limited. Resilience in this context cannot be based on imported or generic models. It needs to be embodied, contextualized, and acutely aware of the fine-grained dynamics of place. For this reason, this chapter and the book as a whole look for resilience not

A. Cheshmehzangi, *Thailand's Tropical Thresholds*, Urban Sustainability,
https://doi.org/10.1007/978-981-95-4975-7_1

only in structures and systems but also in thresholds—the times and locations where innovation becomes unavoidable and adaptation becomes necessary.

Thailand is experiencing a number of overlapping transitions at the national level, including the shift from rural to coastal concentration, from agrarian to urban livelihoods, and from a comparatively stable climate regime to one characterized by intensification and unpredictability. Although Bangkok frequently takes center stage in discussions about Thai urbanization and climate vulnerability, this book purposefully directs attention to southern and eastern Thailand, which are as important despite being underrepresented in urban studies. Erosion, storm surges, and the loss of mangroves are putting increasing strain on the southern beaches, which stretch from Surat Thani to Songkhla. Rapid industrialization in eastern towns like Chanthaburi and Rayong is putting additional pressure on coastal ecologies and freshwater systems (Gilfillan et al., 2019). These geographies are diverse, yet they share a common challenge: navigating the narrow and shifting terrain between environmental risk and development ambition.

Tropical thresholds are framed both spatially and temporally. The effects of climate change are gradual and frequently invisible until a crisis forces them to be acknowledged. Building resilience before a disaster occurs is the proactive approach recommended by the Sendai Framework for Disaster Risk Reduction (2015–2030). Climate adaptation is also central to SDG 11, which aims to create inclusive, safe, resilient, and sustainable cities. Furthermore, the IPCC emphasizes that coastal and tropical areas are particularly vulnerable to the compounding consequences of global warming. These international frameworks serve as a foundation for national and local action but converting their ideas into workable plans calls for multi-scalar thinking, grounded techniques, and contextual knowledge.

In order to achieve this, this book uses an integrated methodological approach that combines governance assessment, participatory fieldwork, and geospatial analysis. Spatial patterns of risk, such as urban heat islands, flood-prone areas, and deforestation trajectories, can be found with the use of geospatial techniques. Community knowledge, local stories, and lived experiences of change are captured through participatory fieldwork, especially in places that top-down planning frequently ignores. Institutional capabilities, policy alignments, and the vertical and horizontal coordination mechanisms required for adaptive planning are all examined in governance assessments. By integrating these techniques, the book goes beyond identifying weaknesses to reveal the relational dynamics that either facilitate or obstruct change.

The book's main contention is that three factors environmentally sensitive, culturally distinctive, and institutionally integrative, are necessary for sustainable climate resilience in tropical coastal communities. To be environmentally sensitive, one must embrace water, heat, and biodiversity as design partners rather than enemies and work with the dynamics of the tropics rather than against them. Cultural distinctiveness guarantees that adaptation tactics are in line with regional customs, values, and knowledge bases. By bridging the gaps between local customs, community-based initiatives, and national mandates, institutional integration produces more robust and cohesive solutions.

Thus, this first chapter is a conceptual as well as a cartographic exercise. It charts the landscape of tropical resilience, not only geographically but also conceptually, conceptually, and methodologically. It poses the following question: What kinds of futures are possible on the other side of the thresholds being crossed? It recognizes that being resilient means not just surviving a changing climate but also embracing it and moving forward into new paradigms.

The reader will come across case studies from peri-urban and coastal communities, planning interventions, traditional practices, scientific findings, and local wisdom as we progress through the ensuing chapters. The goal is to highlight various routes toward regenerative and climate-positive futures in the tropics rather than provide a single blueprint. The southern and eastern thresholds of Thailand are not destinations; rather, they are points of entry into a new, emergent, contextualized, and profoundly alive climate consciousness.

Finally, the theoretical and practical underpinnings of the book are established in this introductory chapter. It places Thailand's urban and coastal changes in the larger framework of tropical environmental thresholds, fast urbanization, and climate change. It presents the idea of "tropical thresholds," which are climate tipping points unique to hot, humid, and water-rich regions, by drawing on important global and regional frameworks, including the IPCC, SDG 11, and the Sendai Framework (de Bruin et al., 2018). In addition to explaining the integrative methodology that combined geospatial analysis, participatory fieldwork, and governance assessment, the chapter describes the spatial scope (southern and eastern Thailand). It also lays out the main argument of the book, which is that cross-scale institutional action, cultural specificity, and environmental sensitivity must all be balanced for resilience-building in tropical coastal cities to be successful.

1.1.1 Tropical Geographies and Environmental Fragility

Thailand's tropical identity is more than just a description of climate; it is a dynamic, living state that influences how people create their cities, engage with their surroundings, and adapt to change. This is particularly noticeable in the country's eastern and southern regions, where tropical geographies collide with escalating climate stresses. These zones, which are characterized by dense ecological systems, long coastline margins, significant rainfall, and persistent humidity, are becoming more and more vulnerable to environmental change. Contextualizing the newly emerging "tropical thresholds" that this book introduces requires an understanding of these distinct geographies.

The southern region is a complex coastal region that stretches along the Gulf of Thailand and the Andaman Sea (Bennett et al., 2015; Mahattanachai et al., 2021; Saramul & Ezer, 2014). River deltas, mangrove forests, and complex estuary systems that sustain ecological variety and human livelihoods can all be found there. The dual nature of the region—being both environmentally rich and exposed to the climate—is exemplified by provinces such as Songkhla and Nakhon Si Thammarat. The southern region experiences nearly continual hydrological stress due to temperatures that frequently exceed 35 °C and yearly rainfall that exceeds 2000 mm in several places. Recurrent issues include coastal erosion, tidal surges, and flash flooding. Rapid urbanization, aquaculture development, and tourism-driven growth all increase these environmental concerns by causing land-use change, deforestation, and ecosystem disruption.

To the east, regions such as Chanthaburi, Rayong, and Trat reveal a different but equally precarious landscape. These provinces, which are situated nearer the Eastern Economic Corridor's (EEC) industrial corridors and marine trade routes, are changing as a result of port development, infrastructure investment, and economic expansion. However, because of their location, they are vulnerable to rising sea levels, erratic rainfall patterns, and saltwater intrusion. Global climate change is increasingly upsetting the monsoonal rhythm that these eastern provinces experience, which consists of a distinct dry season followed by periods of heavy rainfall. As a result, both urban infrastructure and rural agriculture are put under stress by a cycle of water extremes that includes drought and deluge.

Living on ecological boundaries is exemplified by southern and eastern Thailand, which are regions where infrastructure and nature, heat and water, and land and sea meet. Although these areas have historically adapted to tropical rhythms, they are currently dealing with change on a larger scale and at a faster rate than in the past. Crucially, these shifts aren't happening alone. They reflect larger worldwide patterns of climate instability, which include a breach of historical climate standards, an increase in heatwave frequency, and stronger storms. What used to be predictable is suddenly unpredictable.

Intersecting stresses exacerbate these regions' tropical vulnerability even further. In many coastal areas, land subsidence—a result of excessive groundwater extraction—is speeding up the relative rise in sea level. Low-lying communities are now more vulnerable to storm surges as a result of the loss of mangroves and other natural barriers. In the meantime, floodplains, wetlands, and forested hillsides are being encroached upon by urban sprawl, which is frequently unplanned and informal. In addition to increasing exposure to risks, this spatial expansion lowers the ability of ecosystems that formerly buffered those risks to adapt.

However, the vulnerability extends beyond the surroundings. It is both infrastructure-based and institutional. The ability or coordination systems to adequately address climate threats are lacking in many southern and eastern towns. Reactive planning, fragmented governance, and a lack of climate data have caused gaps in resilience-building initiatives. In this situation, climate change not only increases the threat but also reveals vulnerabilities by highlighting the flaws in governance structures, infrastructure networks, and planning systems.

Importantly, in the narrative of climate change, the tropics are not helpless victims. They are hubs for biological productivity, human adaptation, and high energy exchange. However, the same characteristics that distinguish tropical regions also make them more susceptible to change. For instance, in regions that are already hot and humid, a mere 1.5 °C increase in the average temperature might have disproportionately negative impacts on ecosystem function, production, and public health

(Cheshmehzangi, 2025). Heat stress is becoming a greater concern in both urban and rural areas as a result of the combination of high baseline temperatures and high moisture content, which strain human thermoregulation.

Therefore, exposure is not the only factor contributing to these regions' fragility; thresholds are also being breached. Coral bleaching in warmer waters, fishery collapse from ocean acidification, and biodiversity loss from land conversion are examples of ecological tipping points that are becoming more apparent. Social thresholds, on the other hand, are equally important: when communities can no longer depend on seasonal patterns, when adaptation costs surpass local resources, and when conventional livelihoods become unsustainable. Adaptation becomes existential in these situations and is no longer optional.

The geographic and environmental framework for comprehending Thailand's tropical thresholds has been established in this section. As the book goes on, we'll see how these thresholds are felt locally in planning strategies, infrastructure design, municipal responses, and community resilience activities. The southern and eastern regions serve as living laboratory for climate resilience in addition to being risk areas. Paradoxically, their vulnerability provides insight: a prism through which the rest of the world might comprehend what it means to live—and adjust—at the cusp of change.

1.2 From Global Commitments to Regional Realities

This section places Thailand in the context of global frameworks like the Paris Agreement, SDG 11, the Sendai Framework, and IPCC reports. It examines the ways in which the experienced urban and ecological realities of tropical regions interact with these global narratives. The global conversation on disaster risk reduction, climate change, and sustainable urban development has been influenced by a number of international frameworks in recent decades. A common language for discussing planetary risks is provided by documents like the Paris Agreement, the Sustainable Development Goals (SDGs), especially SDG 11, the Sendai Framework for Disaster Risk Reduction, and the reports of the Intergovernmental Panel on Climate Change (IPCC). These frameworks encourage coordinated action in the face of global environmental crises by providing nation-states with direction, urgency, and normative benchmarks. However, these commitments must face the challenging terrain of regional specificity—of tropical climates, ecological sensitivities, governance constraints, and cultural contexts—when they cascade downward into particular geographies, such as southern and eastern Thailand.

This section examines the frequently uncomfortable interplay between these global obligations and local reality. It highlights the necessity for contextual, flexible, and pluralistic answers while criticizing the propensity to universalize climate and resilience solutions. Examining how global frameworks' tenets are understood, translated, and occasionally opposed at sub-national levels, particularly in the tropics, is the aim, not discounting them.

1.2.1 The IPCC and the Limits of Generalization

The most reputable scientific consensus on climate change is represented by the IPCC's assessments. Although their estimates and models are global in scale, their conclusions have a significant impact on local and national policies. As a member of the global south, Thailand is frequently noted as being extremely sensitive to climate-related hazards, such as rising sea levels, intense heat, coastal flooding, and disturbances in water cycles. The IPCC has highlighted in previous reports that tropical and coastal regions are disproportionately at risk from the compounding effects of urbanization, socioeconomic disparity, and warming.

Although localized vulnerability is acknowledged in these publications, local decision-making is frequently not informed by the coarse resolution of global climate models. Aggregated risk categories fall short in capturing the realities of southern Thai cities where land subsidence, storm surges, and inland flooding converge. Furthermore, in prevailing climate narratives, desertification or glacial melt are given more emphasis than the biophysical features of hot-humid settings, such as high wet-bulb temperatures or persistent humidity stress. Thus, there is still a gap between macro-scale projections and micro-scale lived experience, even though the IPCC's work offers a crucial body of evidence.

Important questions are brought up by this: How might the knowledge of the global climate improve the planning of tropical cities? How might indigenous knowledge, culturally entrenched resilience strategies, and hyperlocal data be incorporated into these global assessments? These are not just technical issues; they also relate to the need to refocus the global climate agenda toward more diverse geographies and modes of knowing, as well as epistemic justice.

1.2.2 SDG 11 and the Tropics: Whose Urban Future?

Urban transformation is directly addressed by SDG 11—"Make cities and human settlements inclusive, safe, resilient, and sustainable"—one of the 17 Sustainable Development Goals (Allam & Cheshmehzangi, 2024). It focuses on enhancing planning, resilience, housing, and transportation. However, a northern or temperate bias is frequently reflected in its indications. Typically, urban systems that presume a certain degree of institutional capability, governance coherence, and infrastructure standardization frame resilience criteria including building codes, emergency readiness, and infrastructure durability.

The hybrid urbanisms of many towns and peri-urban areas in southern and eastern Thailand, on the other hand, combine modern technologies and vernacular practices, formal and informal institutions, centralized policies, and community-based adaptation (Cheshmehzangi et al., 2023). These locations do not readily fit into worldwide urban development models due to their physical characteristics, which include stilt houses, floodplain communities, dispersed settlements, and linear growth along coastal roads. Additionally, the technocratic interpretation of resilience as merely sturdy infrastructure is challenged by cultural ties to water, forests, and seasonal cycles.

This does not mean SDG 11 is irrelevant to Thailand. Actually, numerous local efforts have been sparked by its rhetoric regarding inclusive and participatory planning. But to make sure that resilience is about fostering adaptability through social cohesion, ecological consciousness, and institutional flexibility rather than merely hardening cities, it is necessary to contextualize the goals. It is impossible to attain resilience in tropical regions without considering the subtleties of climate-culture-place relations.

1.2.3 The Sendai Framework and Disaster Governance

A comprehensive strategy for lowering disaster losses in terms of lives, livelihoods, and health is provided by the Sendai Framework for Disaster Risk Reduction (2015–2030) (Pearson & Pelling, 2015). Thailand's national disaster strategy mirrors its emphasis on assessing disaster risk, bolstering governance, investing in resilience, and improving preparedness. The nation has implemented integrated flood management, early warning systems, and risk-based zoning, especially in urban areas susceptible to flash floods and typhoons.

However, regional implementation is still dispersed. Local governments frequently lack the financial resources and technical know-how to carry out risk assessments or maintain vital infrastructure. Instead of being the result of consistent proactive investment, planning is often reactive, started in the wake of tragedies. Disaster preparedness is made more difficult by the uneven coordination between national and local actors as well as between agencies.

Furthermore, disasters in tropical areas can occur gradually and cumulatively, showing up as a sequence of ongoing stresses rather than as isolated shocks. Examples of these include rice field salinization, continuous coastline erosion, and protracted droughts. Frameworks that manage cumulative climate deterioration—the insidious violence of environmental degradation that gradually threatens lives and ecosystems—must be included to the Sendai Framework's capabilities for addressing acute hazards.

1.2.4 The Paris Agreement and the Role of Subnational Actors

By encouraging countries to make Nationally Determined Contributions (NDCs), the Paris Agreement set the global goal of keeping warming well below 2 °C. Climate-smart agriculture, adaptation plans, and greenhouse gas reduction are all goals under

Thailand's NDCs. Sub-national implementation is still uneven, though, particularly in tropical areas where local priorities frequently prioritize poverty alleviation and economic growth over climate mitigation.

Thailand's south and east demonstrate the possibilities and difficulties of localization. On the one hand, local governments have tested low-carbon tourism schemes, mangrove restoration projects, and renewable energy projects. However, these initiatives are frequently isolated, donor-driven, and short-lived. In order to acknowledge local governments as co-creators of climate strategies rather than as executors of national mandates, it is necessary to provide them with ongoing financial and technical support.

Additionally, the Paris Agreement's top-down focus on emissions obscures the need for adaptation in areas that are more vulnerable to the issue than those that caused it. Resilience for Thailand's tropical coastal communities is more about safeguarding livelihoods, securing freshwater, and maintaining cultural landscapes than it is about accounting for emissions. These bottom-up realities need to be more accurately reflected in the global climate agenda.

1.2.5 Bridging the Gap: Toward Situated Multi-scalar Action

The conflict between regional realities and global obligations is a call for recalibration rather than a failure of international institutions. This book makes the case that it is impossible to model resilience in tropical Thailand from a distance. It must come to light via communication between disciplines, cultures, and scales. Local activity should be inspired by global frameworks, not dictated by them. Through the prism of location, they need to be translated, interpreted, and recreated.

By doing this, we can create a strategy for climate resilience that is both locally based and internationally conscious—one that respects complexity while aiming for clarity, one that recognizes fragility yet invests in potential. Thailand's southern and eastern regions, which are situated on the brink of ecological and socioeconomic thresholds, are not merely beneficiaries of international agendas. They are places of experimentation, education, and change. Furthermore, the future of tropical resilience will be shaped in these disputed regions.

1.3 Conceptualizing Tropical Thresholds

I wanted this book to be unique to the context of Thailand but also to be relevant to other tropical regions and locales; hence, the idea of "tropical thresholds". The chapter's central conceptual innovation, "tropical thresholds", as regionally distinct tipping points in ecological, infrastructure, and sociopolitical systems, is introduced and expanded upon in this section. It provides theoretical support and elucidates the

Fig. 1.1 Floodgate in designated places for new development zones in eastern coast of Thailand

significance of these thresholds for resilience-building (see Figs. 1.1, 1.2, 1.3, 1.4, 1.5, 1.6, 1.7, and 1.8).

The concept of "tropical thresholds" emerges in this book as a critical and place-based lens through which to understand the converging crises facing climate-vulnerable geographies in Thailand and beyond. Although the term "tipping points" has been used for a long time in environmental science to describe irreversible changes in ecological systems that occur when a threshold is crossed, this framework has frequently been based on biophysical feedback loops like Amazon dieback or polar ice melt or global climatological models. The thresholds that appear more quietly and locally are less well-expressed, especially in tropical regions where social, environmental, and infrastructure systems are intricately linked and particularly vulnerable to climate change. The term "tropical thresholds" is theoretically grounded in this section, which suggests that it be used not only as a descriptive term but also as an interpretive tool to understand regionally specific limits—physical, institutional, and cultural—beyond which adaptive capacity starts to deteriorate and resilience becomes less likely.

Tropical thresholds in southern and eastern Thailand are complicated, cumulative processes of transition rather than isolated events of collapse. The interconnectedness of climate stressors, including as heat, humidity, rainfall variability, and sea level rise, and the ability of urban and rural systems to absorb, adapt to, or change in response to these stressors define these thresholds. They are thresholds of exposure, where saltwater intrusion and rising temperatures make some crops unviable; thresholds of

Fig. 1.2 Patches of land and streams that I worry one day will be very different

Fig. 1.3 Floor prone zones of eastern coast in Thailand, places that one day may become patchy islands of the east

Fig. 1.4 There is sign of progress for Thailand's temporary and permanent infrastructure development across the country

habitability, where heat and humidity levels approach or surpass human thermoregulation limits; and thresholds of governance, where institutions do not coordinate across scales, resulting in delayed or misaligned risk responses.

Tropical thresholds differ from their temperate equivalents not only in the degree of environmental risk but also in how deeply ingrained it is in daily living. Climate variability, ecological cycles, and seasonal rhythms have traditionally influenced building styles, settlement patterns, agricultural systems, and cultural activities in tropical regions. The tropics actively contribute to the creation of space and civilization rather than serving as a backdrop for development. Therefore, when these climates start acting differently—for example, when monsoons become unpredictable, when high seas engulf cities, or when inland heat levels continue to rise above bearable levels—the consequences affect not just ecosystems but also livelihoods, social contracts, and institutional coherence.

Additionally, tropical thresholds have an infrastructure component. The coastal and riverine regions of Thailand are home to several towns and cities that were never built to withstand the extremes of the current environment. Under the strain of recurrent flooding, persistent moisture, or rapid deterioration brought on by heat, infrastructure networks—roads, drainage systems, housing stock, and public spaces—are approaching their functional limitations. Mold-infested schools, deteriorating walkways, failing culverts, and homes raised on stilts that are now too short to avoid rising waves are all examples of these tangible boundaries. Materials that deteriorate

Fig. 1.5 Tropical thresholds are also visible in remaining green spaces of riversides in cities like Chumphon in Southern Thailand

under constant humidity, maintenance schedules that presume drier conditions, and planning codes that are frequently taken from non-tropical contexts all contribute to the fragility of infrastructure in tropical environments (se Fig. 1.9 and its peculiar caption).

But tropical thresholds are not only climatic or infrastructural—they are also socio-political. They are crossed when communities lose faith in institutional processes, when relocation becomes the only option, or when traditional ecological knowledge is no longer sufficient to interpret a rapidly shifting environment. To survive in the tropics, traditional populations in Thailand and many other Southeast Asian countries have long relied on seasonal knowledge, environmental cues, and collective land management techniques. Such knowledge systems may start to crumble as urbanization and climate change increase, not because they are out of date but rather because the natural patterns they depend on are changing. In these situations, a community's ability to adapt can tilt into vulnerability—not because of ignorance or opposition, but rather because the structures that once allowed for resilience have been structurally displaced.

Temporal relevance is also attached to the concept of thresholds. Many tropical thresholds are slow-onset, showing up over years or decades, in contrast to abrupt calamities. Because of their gradual character, they are more difficult to identify, finance, and prioritize within policy frameworks that often reward short-term, visible

Fig. 1.6 Highway infrastructure over natural water bodies in Chumphon is an example of misguided planning

actions. Many tropical communities are experiencing creeping crises as a result of this temporal subtlety, in which the threshold has been crossed but the effects have not yet fully shown. For instance, over a ten-year period, saltwater intrusion may gradually erode freshwater aquifers, leaving a village without drinkable water only after it is too late to stop it. In a similar vein, mangrove forests may gradually deteriorate without anyone noticing until a single storm event exposes the lost protection, which could have disastrous consequences. Reorienting planning and risk management from reactive crisis reaction to anticipatory governance based on early warnings and cross-sectoral collaboration is necessary due to these creeping dynamics.

Thus, conceptualizing tropical thresholds encourages a reinterpretation of resilience as the ability to purposefully, with agency and forethought, cross thresholds rather than the ability to return to a previous condition. Once crossed, some thresholds might not be able to be reversed, while others might present chances for change. The difficulty is in differentiating between hazardous tipping points and fruitful shifts, between collapse and metamorphosis. This entails developing the capacity to identify when a threshold is approaching and to get ready for what awaits planners, legislators, and communities. It also entails recognizing that thresholds are not uniformly applied: a change that is bearable in one metropolis can be disastrous in a nearby fishing hamlet; an issue that is an annoyance in a posh neighbourhood (or wealthy district) could become a disaster in an informal community.

Fig. 1.7 An example of roadside drainage system that my hold much of the flooding just outside the Tha Chana District in Surat Thani which borders Chumphon's southern edge

Ultimately, the idea of tropical thresholds is a conceptual call to urgent, relational, and contextual thinking. It requires us to be aware of boundaries—not as obstacles to be surmounted, but as cues to reconsider how we interact with the environment, institutions, and each other. These boundaries are not theoretical in southern and eastern Thailand. They're being crossed already. The question is not if change is coming, but if it can be greeted with climate-sensitive resilience that is grounded in place and courageous enough to change what no longer works. According to this perspective, tropical thresholds are more than just lines in the sand. Futures are created, uncreated, or remade at these pivotal moments.

1.4 Climate Resilience as Situated Practice

The chapter now turns its attention to the practical and operational aspects of climate resilience. It highlights the necessity of responses that are contextual in terms of culture and climate, particularly in tropical cities that have to strike a balance between ecological constraints and development pressures. The phrase "climate resilience" has proliferated in planning, governance, and development discourse in recent years. However, in spite of its extensive application, the idea is frequently detached from

Fig. 1.8 It pains me to see the delicate tropical thresholds in the outskirts and smaller towns begin to fade; an unsettling reminder of the (urban) transformations that reshaped beautiful cities like Bangkok decades ago

the actual reality of particular locations and remains abstract, technocratic, or unduly homogenized. In tropical cities, where resource scarcity, fast urbanization, and deeply ingrained cultural habits all come together with climate change, this gap is more noticeable than anywhere else. Resilience in southern and eastern Thailand must be viewed as a contextual practice that is responsive to the cycles of tropical life, adapted to climatic conditions, and ingrained in cultural logics rather than as a general goal.

Realizing that location matters is the first step towards developing situated resilience. In addition to physical weaknesses, the tropics' extreme heat, high humidity, seasonal rains, and dense vegetation also influence social behaviours, architectural traditions, and urban structures. Vernacular solutions to these issues have long existed in Thailand, including community water management based on Buddhist and animist customs, open-air ventilation plans, and raised wooden homes for flood protection. However, contemporary planning paradigms that place a higher priority on concrete infrastructure, linear drainage systems, and imported design templates better suited to temperate zones have frequently marginalized these embodied forms of knowledge.

Cities in southern and eastern Thailand are under more and more pressure to fit in with global urban imaginaries and national development objectives as they expand. The ecological knowledge ingrained in conventional settlement patterns is in danger of being overwritten by extensive infrastructure projects, industrial corridors, and real

Fig. 1.9 Sometimes it is difficult to know whether you have been cursed or blessed, and here in Chumphon was one of those moments. The sign warns of trains, yet the crossing feels as if it belongs to another time—narrow, uneven, and barely held together against the weight of daily use. It was eventually closed to cars and larger vehicles; and I assume for obvious good reasons. The infrastructure exists, but it is fragile, exposed to the slow decay of the tropical climate and the pressure of motorbikes and pedestrians who rely on it. What seems like a gateway of movement is also a reminder of vulnerability, where progress and neglect coexist in a delicate balance. My navigation system, when driving through and out of Chumphon, seemed determined to bring me back to the same spot every time. Repeatedly, the route led me to this fragile piece of railway infrastructure, as though insisting that I must cross it. Yet, crossing was impossible. The way was blocked, and so I was forced to turn back, to trace new routes through the city's narrow lanes and congested arteries. I tried different paths, winding through unfamiliar streets, until eventually I surrendered to the map's logic and followed the longer way out—onto the main highways that cut across the urban waterbodies, the slow-moving streams and shaded creeks that crisscross Chumphon's edge. As the city receded behind us, the clamour of traffic gave way to the rising wail of ambulance sirens. They carved their way through the congestion, urgent and unrelenting, heading toward a scene that soon revealed itself before us. A major accident: one vehicle collapsed on the road, others tangled in its aftermath, metal twisted and bodies scattered in chaos. It was a sobering sight. And then came the realisation—had we arrived here 20 min earlier, we might have been part of that wreckage. In that moment, I could not help but see meaning in the forced detours of Chumphon. The blocked railway crossing, the halted path, the insistence on diversion—what felt like frustration had, in truth, been a quiet guide. A protective barrier, steering us away from danger, holding us back just long enough to keep us safe. Perhaps it was the city's infrastructure, imperfect and fragile though it may be, that shielded us. Or perhaps it was the land itself, and the spirit of Thailand woven into its roads and crossings. Either way, what first felt like a curse revealed itself as a blessing: an unplanned delay that became, in retrospect, a silent guardian. Thank you Chumphon, and thank you Thailand!

estate growth. In the meantime, marginalized communities—farmers, fishermen, and informal settlers—are left to deal with worsening heat, diminishing groundwater quality, and rising sea levels without much institutional assistance. The need for located resilience is especially critical in this conflict between top-down aspiration and bottom-up fragility.

More is needed to operationalize climate resilience in the tropics than just flood or hazard mapping. It entails comprehending how individuals adjust to their unique climatic, cultural, and economic circumstances. Adaptation is already under place in many Thai coastal communities—not through official plans, but through experimentation, improvisation, and routines. Examples include community-built flood barriers, rain-fed aquaculture, seasonal relocation, floating gardens, and forestry initiatives spearheaded by local monks. Although these actions might not easily fall into one of the policy categories, they are concrete, culturally relevant reactions to changing climate conditions.

Crucially, the issue of scale must be addressed by located resilience as well. Local activities are frequently embedded inside multi-scalar risk and governance systems in tropical climates. Even while a village can reestablish a mangrove barrier, upstream development or sea level rise brought on by global emissions may make it less effective. In order to respect local agency and improve systemic coherence, localized initiatives must be connected to regional strategies, national policies, and international frameworks. This is especially true in Thailand, where decentralization has sparked municipal creativity but also resulted in fragmented coordination and unequal capacities.

The function of social capital and collective memory is another essential component of situated resilience. Historical memory of previous floods, droughts, or disasters can be a valuable resource in tropical environments, where environmental risks are cyclical and frequently cumulative. In times of crisis, communities with robust horizontal networks—such as cooperatives, schools, informal clubs, and temples—tend to mobilize more successfully. Additionally, these networks function as forums for experimentation and information sharing, allowing place-based solutions to develop naturally.

Rethinking professional roles and planning tools is another requirement of climate resilience as a contextual practice. It is imperative that planners, engineers, and policymakers use participative, iterative, and narratively sensitive approaches instead of relying solely on universal measurements. Designing with water, using vegetation to cool, or acknowledging seasonal mobility as a valid planning feature are some examples of how to work with rather than against climatic fluctuation in a tropical setting. Redefining infrastructure may also entail including mangroves, ponds, cultural customs, and trust in addition to roads and buildings.

Ultimately, resilience in tropical Thailand needs to be based on attunement—to the environment, to the culture, and to the people who live there—rather than abstraction. Progress must be redefined in ways that are inclusive, context-sensitive, and forward-looking in order to strike a balance between development and ecological limits. Situated resilience is a process of ongoing negotiation—between the old and the new, local and global, risk and possibility—rather than a destination. Tropical cities can become regenerative—cities that not only endure in a changing climate but also change with it, according to their own terms—by embracing this complexity.

1.5 Methodological Foundations and Integrative Approach

An integrative methodological approach—one that crosses disciplines, scales, and knowledge systems—is necessary due to the complexity of climate resilience in tropical regions like southern and eastern Thailand. The triangulated framework used in this book is based on three methodological pillars: governance assessment, participatory fieldwork, and geospatial analysis. A multi-layered understanding of tropical thresholds and context-specific resilience practices is made possible by the combination of the unique insights that each pillar provides.

The empirical foundation of this work is geospatial analysis, which offers spatial proof of exposure to infrastructure, urbanization, and environmental stress. To find patterns of vulnerability and ecological transition zones, satellite imagery, GIS-based flood mapping, land use change detection, and thermal surface analysis were used. At several stages, from the regional to the local level, these spatial insights aid in identifying climate thresholds, such as the expansion of coastal vegetation, the incursion into floodplains, or the growing urban heat pockets.

This spatial data is enhanced by participatory fieldwork, which puts lived experience front and center. Interviews, focus groups, and participatory mapping activities were used during field trips in six provinces to involve local communities, municipal officials, religious leaders, and youth organizations. This approach shed light on the everyday realities of climate stress and exposed community-driven innovations, informal adaptation strategies, and vernacular knowledge that are frequently disregarded in formal planning processes. Additionally, it offered insightful information about the sociocultural aspects of resilience, such as local priorities, collective memory, and trust networks.

The third methodological layer was governance assessment, which concentrated on cross-sectoral cooperation, institutional capacities, and regulatory frameworks. This approach assessed how national policies, regional plans, and local activities align—or do not align—in addressing climate resilience through policy studies, government official interviews, and institutional mapping. The issues of vertical integration, budgetary autonomy, and decentralization dynamics that influence adaptive action in Thailand's tropical cities received special attention.

This work embraces methodological pluralism by integrating these three foundations. Instead of making simplistic diagnoses, it combines institutional logics, human narratives, and spatial patterns. This integrated approach guarantees that resilience is viewed as a social and place-based process, rooted in both ecological reality and community ambitions, rather than just a technical obstacle. The book's central argument—that establishing resilience in tropical environments necessitates integrating institutional coherence, cultural distinctiveness, and environmental sensitivity into a single, cohesive, flexible, and context-responsive framework—is thus reflected in the methodological design.

1.5.1 Merging Data, Dialogue, and Design

This section describes the book's multi-method approach, which includes participatory fieldwork to record local narratives, governance analysis to comprehend institutional enablers and constraints, and geographic mapping to uncover risk and change trends. The work has also benefitted extensively from fieldwork in 2023 and 2024, research exchanges through our multiple stakeholder engagements, and site visits and on-site recordings.

Using a multi-method research approach that purposefully combines data, discussion, and design, this book aims to comprehend and promote climate resilience in Thailand's tropical regions. The intricacy of climate thresholds, adaptive capacity, and place-based responses in tropical ecosystems cannot be sufficiently captured by a single method, as this integrative approach recognizes. By combining community knowledge, institutional insights, and geospatial evidence, this study creates a multi-layered understanding that is both contextually grounded and analytically sound. The logic and application of this triadic methodology are explained in detail in this section, emphasizing how each method offers unique insights and how they work in tandem to inform the design of futures that are climate resilient.

Geospatial mapping, the first methodological pillar, was used to display and measure the patterns of land use change, infrastructure vulnerability, and environmental stress in the research locations. The study identified significant hotspots of risk, including urban areas encroaching on floodplains, towns at risk from coastal erosion and storm surges, and locations experiencing intensifying urban heat, using techniques like satellite images, GIS-based overlays, elevation models, and thermal imaging. Beyond just charting exposure, these spatial analyses showed where thresholds were being crossed over time, frequently in subtle and cumulative ways. Thermal imaging, for example, revealed a notable rise in surface heat in peri-urban areas that were formerly dominated by wetlands and orchards in Nakhon Si Thammarat. To identify high-risk urban expansion regions in Rayong, coastline recession rates were calculated with development density. Such information is used as a diagnostic tool to direct responsive activities in addition to providing proof of change.

Spatial data can show trends, but it is unable to convey the complexity of local adaptation or the breadth of lived experience. Participatory fieldwork, the second methodological pillar, is essential in this situation. A wide range of actors participated in field engagements throughout southern and eastern Thailand, including local government representatives, religious leaders, and community elders (Buchenrieder et al., 2017). The study recorded how communities view climate risks, remember past adaptations, and imagine resilient futures through focus groups, participatory mapping sessions, semi-structured interviews, and ethnographic observation. These discussions brought to light information that is frequently hidden in official records, like the seasonal adaptations of nomadic fishing groups, the usage of Buddhist temples as refuge during floods, or the contribution of women's cooperatives to innovative water storage. Co-creation of spatial information was also made possible by participatory tools: community members contributed their own mental maps of risk and resource distribution, challenged presumptions in external data, and improved hazard maps. Rather than being purely extractive, these interactions were presented as mutually beneficial ones in which research served as a platform for empowerment, acknowledgment, and shared learning.

Understanding the institutional environment influencing resilience action was the goal of the third pillar, governance analysis. The study outlined the characteristics of climate governance in Thailand by reviewing national policies, regional plans, and municipal development frameworks and conducting interviews with officials at all levels. A patchwork of systemic limitations and enabling initiatives resulted. On the one hand, a greater understanding of climate risk is reflected in national measures like the National Adaptation Plan and the environmental protections of the Eastern Economic Corridor. Conversely, municipal governments frequently deal with disjointed mandates, little financial independence, and uneven collaboration with federal organizations. Institutional barriers, including jurisdictional ambiguity or budget release delays, frequently hindered resilience just as much as physical threats. Through the identification of champions within municipal offices, unofficial departmental alliances, and chances to incorporate resilience into zoning and development approval procedures, governance research exposed both the gaps and the opportunity for systemic change.

This triadic method's key component is not just the use of many techniques but also the intentional blending of their results. Design, data, and conversation were all viewed as overlapping components rather than as distinct streams. For instance, to validate, challenge, and enhance quantitative insights, geospatial findings were introduced into community workshops. In turn, community narratives shaped the interpretation of spatial data, making sure that mobility patterns, cultural attachments, and social networks were taken into account in addition to topography and infrastructure when defining vulnerability. In a similar vein, institutional interviews were examined in light of spatial exposure maps and community experiences, which assisted in placing governance inadequacies in the context of actual geographies and daily stresses. This synthesis made it possible to move from discrete diagnostics to integrated design thinking, where resilience tactics could be adapted to the local environment, guided by multi-scalar understanding, and incorporated into practice and policy.

The book's central epistemological position—that resilience is relational, co-produced across knowledge systems, and inextricably linked to place—is reflected in the integration of data, discourse, and design. It cannot be completely modelled in software or enforced through top-down architecture. Instead, it needs to be developed in consultation with communities, validated by data, and backed by enabling governance. With Thailand's tropical urbanism, this technique is therefore more than just a study approach; it is a political and ethical commitment to multiple knowledge, grounded action, and the co-creation of climate-positive futures.

1.6 Toward a Regenerative Urban-Climate Nexus

There is no denying the need for a paradigm shift as Thailand's eastern and southern provinces prepare for significant urban and climatic changes. According to this book, the path forward involves intentional development of a regenerative urban-climate nexus—a dynamic connectivity between ecosystems, communities, and institutions that permits not only survival but renewal—rather than merely adaptation or risk mitigation. In this sense, regeneration refers to more than just environmental recovery. It highlights how cities and their people can mend broken ties to the natural world, revive cultural traditions, and rethink government structures that are appropriate for a time of growing climate uncertainty.

Three interconnected principles—institutional action, cultural distinctiveness, and environmental sensitivity—are at the heart of this regenerative vision. While each is necessary, none is adequate on its own. Restoring degraded ecological buffers like mangroves and wetlands, adopting heat-moderating vernacular design, and acknowledging water as a life-sustaining force rather than just a threat are all necessary components of environmental sensitivity. Cultural specificity guarantees that initiatives to enhance resilience are grounded in regional social norms, collective identities, and knowledge systems. Cultural patterns offer the framework for significant and long-lasting change, whether through communal water rituals, faith-based forest management, or the adaptive reuse of traditional structures. Finally, institutional action gives these initiatives coherence and scalability. Without policy support, budgetary commitment, and effective coordination, even the most innovative community responses risk remaining isolated or short-lived.

A regenerative nexus recognizes that climate resilience is a dynamic process that is negotiated over time, across scales, and amongst players rather than a technological formula. Instead of a one-way influence, it reframes the relationship between cities and climate as one of mutual moulding. Cities' infrastructure, pollution, and land usage all affect the climate. Cities are reshaped by the climate as a result of stress, disruption, and disturbance. Instead of trying to eliminate this contact, a regenerative strategy aims to change it, transforming vulnerable areas into places for experimentation, education, and hope.

This work is reflected and expanded upon in the following chapters. Beginning with cases that trace ecosystem degradation and regeneration across Thailand's tropical beaches, the first few chapters delve deeply into the topic of environmental sensitivity. These chapters show how ecological systems can recover from injury if given the time, space, and attention they need. They also demonstrate how rapidly increasing urbanization and mismatched infrastructure might unintentionally set off thresholds, converting controllable stress into irreparable loss.

Place-based resilience is built on cultural distinctiveness, which is examined in the middle chapters. They draw attention to indigenous customs, community-led projects, and socio-spatial trends that influence adaptation efforts on the ground. From peri-urban populations' flood-adapted livelihoods to conservation efforts influenced by Buddhism, these chapters demonstrate how culture can be a resource and a guide for dealing with environmental unpredictability. We then delve into a case of urban sprawl and expansion as well as coastal vulnerabilities before concluding the book. The last chapters examine the function of planning, multilevel coordination, and governance as they relate to institutional action. In particular, an emphasis is given to coastal management issues as well as new development zones, such as coastal urban corridors. In addition to highlighting the discrepancies between national mandates and local realities, they examine the opportunities and gaps in Thailand's decentralized administrative landscape and present new examples of cooperative, flexible governance. These chapters demonstrate how institutions have the power to either limit or promote regenerative resilience through innovative policy, boundary-spanning leadership, and experimental planning tools.

These three components don't work together to provide a comprehensive strategy or a set of rules. Rather, they offer a variety of strategies, all based on empirical data, locally grounded, and motivated by a common goal of creating futures that are not only sustainable but also just and healing. The book provides a counter-narrative to climate fatalism by doing this. It implies that there is both risk and opportunity in the humid heat and tidal flux of Thailand's tropical cities—for institutions that serve rather than govern, for new urban imaginaries, and for rediscovered communal wisdoms.

We encourage readers, planners, academics, students, and practitioners, to examine this regenerative nexus in the upcoming chapters, not as a far-off objective but as a framework that is already beginning to take shape in various initiatives and fragments throughout Thailand's resilient yet vulnerable landscapes. The benchmark has been surpassed. Whether we react with care, bravery, and the resolve to regenerate, or with fear and fragmentation, will determine what happens next.

References

Allam, Z., & Cheshmehzangi, A. (2024). *Sustainable futures and green new deals* (pp. 1–137). Palgrave Macmillan.

Bennett, N. J., Dearden, P., & Peredo, A. M. (2015). Vulnerability to multiple stressors in coastal communities: A study of the Andaman Coast of Thailand. *Climate and Development, 7*(2), 124–141.

Buchenrieder, G., Dufhues, T., Theesfeld, I., & Nuchanata, M. (2017). Participatory local governance and cultural practices in Thailand. *Cogent Social Sciences, 3*(1), 1338331.

Cheshmehzangi, A. (2025). The importance of health-sustainability-peace nexus in making resilient cities. *Peace and Sustainability, 1*(2), Article 100012.

Cheshmehzangi, A., He, B. J., Sharifi, A., & Matzarakis, A. (2023). Climate change, cities, and the importance of cooling strategies, practices, and policies. In *Climate Change and Cooling Cities* (pp. 2–19). Springer Nature Singapore.

de Bruin, S., Knoop, J., Visser, H., & Ligtvoet, W. (2018). *Linking water security threats to conflict: An Exploration of Pathways*, PBL Netherlands Environmental Assessment Agency, The Hague. PBL publication number: 3039, Report Publication.

Gilfillan, D. A., Sawdon, J., Harpham, K., Vinh, M. K., Cooper, R., Mather, R., Buaban, S., Chuangcham, S., Putchakarn, S., & Srinate, V. (2019). *Sea of the Rayong provincial development plan-sustainable development pathway report.* https://sea.nesdc.go.th/th/wp-content/uploads/2024/07/4sea-report-7-4.pdf Accessed on 22 August 2025.

Mahattanachai, T., Morley, C. K., Charusiri, P., & Kanjanapayont, P. (2021). The Andaman Basin Central Fault Zone, Andaman Sea: Characteristics of a major deepwater strike-slip fault system in a polyphase rift. *Marine and Petroleum Geology, 128*, Article 104997.

Pearson, L., & Pelling, M. (2015). The UN Sendai framework for disaster risk reduction 2015–2030: Negotiation process and prospects for science and practice. *Journal of Extreme Events, 2*(01), 1571001.

Saramul, S., & Ezer, T. (2014). Spatial variations of sea level along the coast of Thailand: Impacts of extreme land subsidence, earthquakes and the seasonal monsoon. *Global and Planetary Change, 122*, 70–81.

Srinivasan, G., Agarwal, A., & Bandara, U. (2024). Climate change impacts on water resources and agriculture in Southeast Asia with a focus on Thailand, Myanmar, and Cambodia. In *The Role of Tropics in Climate Change* (pp. 17–32). Elsevier.

Chapter 2
Coastal Vulnerabilities and Hydro-meteorological Risks: Exposure to Coastal Hazards in Krabi and Nakhon Si Thammarat

2.1 Mapping Exposure: Coastal Geographies and Climatic Drivers

Thailand's southern shore is a region of increasing vulnerability as well as an important biological resource. Provinces like Krabi and Nakhon Si Thammarat, which stretch along the Andaman Sea and the Gulf of Thailand, are becoming more vulnerable to hydro-meteorological risks as a result of a confluence of changing coastal geomorphology, sea level rise, and climatic variability. An overview of the meteorological and physical factors influencing various coastal systems is given in this section. By outlining the geographic circumstances that render particular regions especially vulnerable to environmental risk and charting exposure zones, it prepares the ground for more thorough investigation.

Different yet complementary coastal geographies are represented by the twin provinces of Krabi and Nakhon Si Thammarat. The Andaman Sea island chain of Krabi is well-known for its sandy beaches, karst limestone cliffs, and island chains, many of which are popular tourist sites. Seasonal monsoons, tidal currents, and sporadic storm occurrences have all influenced its extremely dynamic coastal morphology (Curoy et al., 2022a). On the other hand, Nakhon Si Thammarat is located on the Gulf of Thailand and is characterized by vast mangrove belts, low-lying deltaic plains, and agricultural hinterlands that are primarily dependent on coastal and riverine ecosystems. Each province's biological traits and susceptibility to certain coastal dangers are influenced by these variations in geomorphology.

In this context, combining geophysical data with climate patterns is necessary for mapping exposure. Southern Thailand is seeing more unpredictable and severe weather patterns than in the past, according to longitudinal meteorological data that includes temperature changes, rainfall variability, wind intensity, and storm frequency. Because of longer dry spells and more frequent cloudburst events, seasonal monsoons are now less predictable. In the meantime, the frequency and intensity of storm surges linked to cyclonic activity and tropical depressions in the Andaman and

A. Cheshmehzangi, *Thailand's Tropical Thresholds*, Urban Sustainability,
https://doi.org/10.1007/978-981-95-4975-7_2

South China Seas are rising (Suphawajruksakul, 2005). This has made Nakhon Si Thammarat more vulnerable to backwater flooding and coastal inundation because it is situated in the path of numerous storms that originate in the Gulf. On the other hand, Krabi's barrier islands and coastal resorts are becoming more vulnerable to tidal surges.

One important sign of these changes is now coastal erosion. Over the past 20 years, the shoreline of Nakhon Si Thammarat has retreated by up to 30 m in some places, especially where mangrove forests have been cut down for construction or aquaculture. Although less linear, erosion in Krabi is nevertheless concerning, especially around the eastern borders of well-known beaches where the shoreline is reshaped every year by storm-driven waves. Due to their direct effects on local infrastructure, property, and livelihoods, these losses are not only socioeconomic but also environmental.

The incursion of saltwater into freshwater systems and agricultural fields, or salinization, is another crucial factor. Rice fields and groundwater supplies are under danger in Nakhon Si Thammarat's deltaic regions due to an increase in soil salinity brought on by a combination of increased tidal pressure and decreased upstream freshwater flow. Salinization reduces biodiversity and erodes natural barriers against storm surges, which impacts wetland ecosystem health as well as the sustainability of agricultural productivity.

It becomes more evident where thresholds are being passed when these drivers—erosion, storm surge zones, salinity intrusion corridors, and high-risk coastline stretches—are spatially mapped. Clusters of compounding risk are seen in the geospatial overlays created for this study, especially in neighbourhoods constructed on former marsh regions or along eroding beaches. These include peri-urban tourist areas in Krabi that are susceptible to coastal flooding during periods of high tide and intense precipitation. Whole village clusters in deltaic depressions in Nakhon Si Thammarat are susceptible to seasonal flooding, which is growing increasingly common and unpredictable.

Developing context-sensitive resilience measures requires an understanding of these coastal geographies and climate variables. Natural landforms, human modifications to ecosystems, and the buildup of climatic stresses all influence exposure, which is not constant. The spatialized foundation provided by the mapping done in this chapter allows for the layering of adaptive treatments. It also shows how exposure is dynamic and influenced by feedback loops between human development and environmental deterioration. Mapping is a political and ecological act that reveals both risk and responsibility in tropical coastal locations like these. It is not merely a technical task.

By analyzing the morphological and meteorological differences between Krabi and Nakhon Si Thammarat, as well as how these variations result in distinct patterns of risk, exposure, and adaptation requirement, the ensuing subsection will further break down these discoveries.

2.1.1 *Coastal Morphologies and Meteorological Dynamics in Southern Thailand*

Using coastal geomorphology and long-term meteorological data (rainfall, wind patterns, cyclone paths), this subsection presents the physical and climatic landscape of Krabi and Nakhon Si Thammarat (Cheewinsiriwat et al., 2024). It explains how each area/region reacts to hydro-meteorological stressors such sea-level rise, tidal surges, and coastal erosion by drawing a comparison between the sandy beaches and rocky shorelines of Krabi and the mangrove-dominated deltaic systems of Nakhon Si Thammarat.

A thorough analysis of the region's physical characteristics and climatic patterns is necessary to comprehend coastal vulnerability in southern Thailand. The two focus provinces, Krabi and Nakhon Si Thammarat, present a striking contrast in the ways that hydro-meteorological stressors shape and affect tropical coastal systems (Curoy et al., 2022b). A window into the various mechanisms of risk accumulation over Thailand's southern seaboard is provided by their disparate geographies, sediment compositions, shoreline dynamics, and climate exposures.

The rocky and diverse coastline of Krabi, which is situated along the Andaman Sea, is made up of narrow coastal plains, offshore islands, sandy coves, and karst limestone formations (Scheffers et al., 2012). Despite its aesthetic value and economic significance to the tourism industry, the famous limestone cliffs and sea stacks are not naturally resistant to storm surges or sea level rise. Because a large portion of Krabi's coastline is sandy and devoid of silt, its beaches are extremely dynamic and vulnerable to both natural erosion and damage brought on by humans. Due to their frequent lack of built buffers or significant backshore vegetation, the beaches are more susceptible to seasonal storm surges that occur during monsoonal high tides. Frequent tidal flooding events in recent years have damaged resort infrastructure and ruined coastal paths, upsetting ecosystems and livelihoods.

In contrast, the morphological profile of Nakhon Si Thammarat, which is located on the Gulf of Thailand, is significantly different. Due to centuries of sediment deposition and river processes (see Fig. 2.1), the majority of its shoreline is deltaic and mangrove-fringed (Aksornkoae & Bird, 2010; Bidorn et al., 2020). The province contains low-lying agricultural areas, wide intertidal flats, and vast estuaries that have traditionally relied on the protective roles of mangrove ecosystems. Despite its innate ability to disperse tidal energy and filter salinity, the deltaic geomorphology is currently facing mounting pressure from both human activity and climate dynamics. Natural buffer zones have been reduced due to mangrove cutting for aquaculture and coastal development, leaving communities and rice paddies vulnerable to storm surges and long-term salinization. The effects of sea level rise are exacerbated by the delta's soft, muddy substrate, which is especially vulnerable to erosion and land subsidence.

A common exposure to hydro-meteorological variability, influenced by the seasonal monsoons and tropical cyclone systems that dominate much of southern

Fig. 2.1 An example of streams linking to rivers in Nakhon Si Thammarat

Thailand's climate, sits above these morphological distinctions. With its Andaman-facing coastline, Krabi has heavy monsoonal rains from May to October, frequently accompanied by powerful southwesterly winds, according to long-term weather records. These winds speed up shoreline erosion and push tidal surges inland, especially when tidal and precipitation peaks overlap (see Figs. 2.2 and 2.3). Krabi is increasingly experiencing unseasonal storm events linked to regional climate anomalies, despite being less frequently hit by tropical cyclones than Gulf-side provinces. In addition to limiting natural drainage, its insular topography and narrow river outlets cause localized flash floods during periods of intense precipitation.

On the other hand, Nakhon Si Thammarat is directly in the path of typhoons and tropical depressions that originate in the Gulf. Its shore has seen numerous storm landfalls, especially between October and December, according to historical cyclone path research. Both fluvial and coastal flooding are caused by these storms, which frequently have strong winds and a lot of rain. The province's low elevation and level terrain increase the likelihood of protracted flooding, particularly in inland regions where floodwaters have few escape routes. Coastal wave accumulation and shoreline retreat in exposed areas are also influenced by seasonal wind patterns, which are primarily from the northeast during the post-monsoon months. Annual rainfall unpredictability has increased over the last 20 years, according to meteorological

Fig. 2.2 A serene stretch of Krabi's coastline, where calm turquoise waters gently meet a sandy shore framed by lush greenery. In the distance, limestone formations rise subtly against the horizon, embodying the quiet, unspoiled character of this part of the province. The absence of crowds and development here contrasts sharply with Krabi's busier tourist hubs, offering a glimpse into the region's more tranquil and ecologically intact coastal environments

data, with unpredictable dry spells followed by heavy downpours—patterns that put a burden on the resilience of infrastructure and the environment.

The relationship between sea-level rise and land subsidence is another important distinction. Decades of groundwater extraction for aquaculture and agriculture in Nakhon Si Thammarat have caused noticeable sinking in several coastal districts. The effective or "relative" sea-level increase is much greater than the global average when paired with regional sea-level rise, which is currently estimated to be between 3 and 5 mm annually. This decreases the efficiency of drainage and protective systems, increases flood exposure, and speeds up coastline retreat. Even though land subsidence is less severe in Krabi, many shorelines' natural resistance has been compromised by beach compaction brought on by excessive use and poorly managed coastal development.

These variations in weather patterns and coastal morphology highlight the necessity of specialized adaptation measures. Nakhon Si Thammarat has to put more of an emphasis on mangrove restoration, salinity control, and long-term retreat planning in subsiding zones, whereas Krabi might benefit from dune stabilization, zoning changes, and beach replenishment to reduce erosion. Updated flood modelling that takes into account changing storm patterns and sea level projections unique to each province's geographic setting would be advantageous.

Fig. 2.3 The unique characteristics of Krabi's coastal zones reflect a complex interplay of geological formations, rich marine biodiversity, and deeply rooted local livelihoods, creating a landscape where natural beauty and cultural heritage are inextricably linked

In conclusion, Krabi and Nakhon Si Thammarat's coastal landscapes show how exposure to hydro-meteorological risks depends on shape, history, and land use in addition to climate. These landscapes are dynamic systems that are not static; human choices and natural processes will determine how vulnerable they become in the future. Planners and policymakers might shift from reactive to anticipatory resilience—designed in accordance with the terrains they aim to protect—by comprehending these morphologies and climatic patterns in tandem.

2.2 Coastal Hazard Profiles: Erosion, Inundation, and Salinization

A hazard typology for the two provinces is given in this part, along with thorough risk evaluations for tidal flooding, storm surges, acute and chronic erosion, and soil salinization. It illustrates how risk differs across coastal typologies by identifying hazard zones, risk intensities, and recurrence intervals using vulnerability maps and historical event data. A variety of hydro-meteorological risks that are becoming more frequent, intense, and widespread are present in the coastal areas of Krabi and Nakhon Si Thammarat. Storm surges, tidal inundation, coastal erosion, and soil and water resource salinization are some of the most urgent. Both provinces are exposed to these risks, but because of local geomorphology, land use patterns, and the condition of protective ecosystems, the impact's nature, scope, and mechanisms vary greatly. Using vulnerability mapping, historical data, and field-based assessments, this part provides a comprehensive hazard typology for each province, highlighting the distribution of these risks and the insights they provide on long-term coastal fragility.

The most obvious hazard in Krabi is coastal erosion. The province's coastline is dynamic and ever-changing, with more than 150 km of sandy beaches, tiny estuaries, and limestone headlands making up the majority of the coastline. According to vulnerability maps, important beach areas are retreating at rates of one to three meters annually, especially along Ao Nang, Klong Muang, and the eastern coastlines close to Nuea Khlong. Tidal currents are the primary cause of chronic erosion in these regions, which is made worse by sand mining, unchecked coastal expansion, and the clearing of natural vegetation for resort development. Seasonally, during the southwest monsoon, when storm surges coincide with high tides and wave heights rise, acute erosion episodes take place. Protective seawalls have been built in a number of zones in recent years, but they frequently starve downstream sediments, which speeds up erosion in nearby unprotected places.

In Krabi, tidal flooding is usually confined but getting worse, especially in low-lying communities and close to river mouths. The Krabi Estuary's historical tide gauge data indicates an increasing tendency in peak tide levels, with previously safe places now experiencing sporadic inundation during king tides. Hotspots around Ban Thong Nian and Ban Ao Nam, where tidal waves frequently flood highways, farmlands, and informal dwelling clusters, are shown on vulnerability maps. Because of the inadequate elevation and drainage capacity of the infrastructure in these places, tidal encroachment is a persistent annoyance that has long-term effects on public health and property values.

Although salinization is not as common in Krabi as it is in deltaic areas, some locations have seen an increase in soil salinity, especially in brackish water zones next to aquaculture ponds. The combined effects of land cover changes limited freshwater recharge during dry seasons, and seawater intrusion are responsible for this. Although it is not yet a major danger in Krabi, salinization is becoming a threat to mangrove regeneration initiatives and small-scale agriculture, particularly in areas where tidal flows have increased in force because of adjacent dredging or embankment breaches (see Figs. 2.4, 2.5, and 2.6).

Nakhon Si Thammarat, on the other hand, has a deeper and wider profile of hazards, especially in its low-lying eastern districts such Pak Phanang, Hua Sai, and Chian Yai (Attavinijtrakarn, 2019). The deltaic coastline of the province, which is mostly made up of vast mangrove belts and alluvial soils, is extremely susceptible to both acute and chronic erosion. In some areas of Pak Phanang, where land subsidence, mangrove degradation, and weakening sediment flows have combined to cause significant coastline retreat, erosion rates of up to 5 m per year have been observed. According to field observations, whole rows of sheltering casuarina trees

Fig. 2.4 The extensive yet fragile mangrove ecosystems along Krabi's coastline, forming intricate networks of roots that stabilise the shoreline and provide vital nurseries for marine life. These mangroves are a living barrier against erosion and storm surges, yet their delicate balance is increasingly vulnerable to pressures from coastal development, aquaculture, and changing tidal patterns

are now submerged in the ocean, indicating a continuous loss of land that moves inland with each monsoon season.

Due to its direct exposure to storm tracks in the Gulf of Thailand, Nakhon Si Thammarat also has a much higher danger of storm surges. The region is vulnerable to high-impact storm surges, as evidenced by historical cyclone records, including Tropical Storm Harriet in 1962 and Storm Pabuk in 2019. A 20-km-wide area of high exposure, extending from Tha Sala to Pak Phanang, is highlighted on vulnerability maps. This area is where elevation levels fall below 2 m above sea level. These regions are at increased danger because storm surges frequently occur in conjunction with

Fig. 2.5 Krabi's coastal areas remain active with human activities

high rainfall and upstream runoff, resulting in protracted flooding conditions that saturate rural roads, freshwater canals, and rice fields for days at a time.

In Nakhon Si Thammarat, salinization is arguably the most pernicious long-term threat of all the hazards. Electrical conductivity levels have been steadily rising in important agricultural zones, according to soil salinity data gathered over the last ten years. This is especially true during dry seasons when freshwater input is insufficient to prevent tidal invasion. In the Pak Phanang basin, where water control systems have not kept up with subsidence and rising sea levels, this trend is particularly pronounced. Rice fields' productivity has already decreased due to salinization, which has also prompted the switch to salt-tolerant crops like shrimp farming and, in certain situations, caused the abandonment of hitherto productive regions. The effects are not limited to agriculture; they also have an impact on vegetation patterns, household water supplies, and the feasibility of future land use planning.

These varying danger profiles highlight how crucial site-specific planning is. Policy in Krabi has to concentrate on preventing erosion, controlling land use near beaches, and incorporating natural buffers into the growth of the tourism industry. Restoring mangrove ecosystems, funding adaptive water infrastructure, and assisting livelihood transitions in regions where traditional farming is no longer feasible are all urgent and widespread needs in Nakhon Si Thammarat.

Fig. 2.6 A dense belt of mangroves along Krabi's coastline, their tangled aerial roots anchoring the sandy and silty shore while filtering tidal waters. The interwoven root systems create a natural buffer against erosion, absorb wave energy, and provide sheltered breeding grounds for fish, crustaceans, and countless other species. While these mangroves stretch impressively along the coast, their ecological resilience is fragile, threatened by coastal development, pollution, and shifts in salinity driven by climate change. The scene captures both the strength and vulnerability of one of Krabi's most critical natural defences

The fact that these dangers' recurrence intervals are getting shorter and their intensities are rising in both situations/cases indicates that thresholds are being crossed. It is evident from the maps, statistics, and case studies compiled here that climate stress is an emerging present rather than a far-off future. The shorelines and soil chemistry of coastal Thailand are changing, but so are the fundamental underpinnings of its ecosystems, communities, and economy.

2.3 Livelihoods on the Edge: Socio-economic Vulnerability and Exposure

Natural systems that are increasingly under stress are closely linked to the coastal communities of Nakhon Si Thammarat and Krabi. The socioeconomic vulnerabilities of people who live along the coast are becoming more apparent and severe as erosion, flooding, and salinization increase. Although the climatic risks in both provinces are similar, there are notable differences in the livelihood structures, exposure types, and adaptation capacities of communities. These differences are influenced by factors like as economic reliance, geographical settlement patterns, and availability to institutional support.

Small-scale fishing and tourism are the main drivers of Krabi's economy. A service-based economy dependent on beach integrity, aesthetic quality, and coastal infrastructure is supported by the tourism that coastal towns like Ao Nang, Klong Muang, and Koh Lanta draw from both local and foreign visitors. Today, this sector is frequently disrupted by tidal flooding and seasonal erosion. Damaged pathways, flooded beach sections, and increased maintenance expenses are the main causes of resort owners' losses. During high tide or monsoon disturbances, informal workers—such as sellers, boat operators, and cleaners—are disproportionately impacted and frequently lose their daily income. Because many operate outside formal contracts or social safety nets, their livelihoods are highly precarious.

Small-scale fishermen in Krabi's nearshore fishing villages are likewise dealing with cascading stress. Fish spawning grounds are impacted by changes in nearshore salinity and sedimentation, and unpredictable weather patterns have reduced fishing windows. For example, communities in Ban Khao Thong and Ban Ao Luk report greater operating costs and lower catches as a result of more frequent equipment breakage during storm events. Many households are nonetheless at risk because they have little access to credit, insurance, or other sources of income, even with extensive ecological information and adaptive measures like changing the types of capture or moving nets.

The main sources of income in Nakhon Si Thammarat are seasonal fishing, aquaculture, and rice cultivation, which are primarily found in low-lying deltaic regions. These professions are extremely vulnerable to changes in the environment. In districts like Pak Phanang and Hua Sai, recurring salinization events have weakened soil fertility, resulting in lower yields or crop failure. Brackish water shrimp farming is becoming more and more popular among farmers, but it carries significant upfront costs and environmental dangers, especially in places with unpredictable water salinity or inadequate drainage. The gap between those who can adjust economically and those who cannot is growing as a result of the fact that many smallholders lack the funds or land tenure security necessary to make this shift.

Furthermore, both provinces' informal settlements—which are frequently found near canal banks, tidal flats, or reclaimed mangrove lands—are disproportionately vulnerable to erosion and flooding. Field investigations in Nakhon Si Thammarat show that official hazard mapping and infrastructural investments often ignore low-income residents near estuarine zones. Their homes, which are frequently constructed using local materials on stilts, are not well-protected and are more prone to sustain damage during severe weather conditions. Vulnerability cycles are further exacerbated by the slow and frequently self-financed recovery.

It is clear that environmental and socioeconomic pressures are mounting in both locations. In addition to being vulnerable to physical risks, livelihoods are also hampered by a lack of institutional safety nets, inconsistent access to resources for adaptation, and unequal representation in planning procedures. Building significant and equitable climate resilience along Thailand's vulnerable coasts will depend heavily on safeguarding and facilitating livelihoods, especially for informal and small-scale players, as thresholds are continually crossed.

2.3.1 Community Case Studies: Small-Scale Fishers, Rice Farmers, and Tourism Operators

This subsection examines the ways in which hydro-meteorological hazards impact various livelihoods. It highlights the distinct effects on rice farmers in flood-prone deltas in Nakhon Si Thammarat, artisanal fishermen in Krabi, and coastal tourism operators coping with shoreline degradation and tidal encroachment, based on interviews, fieldwork, and local statistics. It talks about the exacerbated vulnerabilities that communities that rely on the stability of natural resources and seasonal cycles face.

In Thailand's coastal regions, livelihoods are closely linked to ecological stability, seasonal cycles, and environmental predictability. The economic foundation of many coastal communities in Krabi and Nakhon Si Thammarat is made up of three main livelihood groups: small-scale fishermen, rice farmers, and coastal tourism operators. These populations, however, are also among the most susceptible to hydro-meteorological risks. This section examines how these communities are coping with, adjusting to, and being limited by escalating climate-related risks, using fieldwork observations, interviews, and local datasets gathered between 2023 and 2024.

Fishing grounds and catch patterns have changed significantly in Krabi, according to small-scale fishermen who work along the Andaman Sea. Due to sedimentation movements and variations in water salinity, artisanal fishermen in villages like Ban Khao Thong and Ban Nai Sa have observed a decrease in nearshore species and a decline in the dependability of traditional fishing zones. These populations are particularly vulnerable to abrupt weather changes and tidal surges since they usually fish from long-tail boats with limited navigation capacity. Interviewees described how unseasonal storms, which have become less predictable and more unpredictable in recent years, had damaged boats and equipment. A One fisher, aged 52, remarked: "*We used to know when the winds would come. Now the sea is confused, and so are we*".

These fishermen's reliance on daily earnings, lack of official insurance, and restricted participation in decision-making procedures all contribute to their vulnerability. Fewer locals are prepared to continue the trade due to the rising risk and diminishing rewards, and many are getting older. As tidal floods and coastline deterioration increasingly impair resort operations, the same climate factors have hindered attempts to diversify revenue, such as seasonal labour in nearby tourism enterprises.

Here are some examples of responses from fishers in the region:

> When I was younger, we could go out every day. The sea had its moods, but we knew them. Now it changes without warning. Some mornings we leave before sunrise, only to turn back halfway because the winds rise too fast. My son doesn't want to fish anymore—he says it's too dangerous, and not worth the trouble. He drives tourists around instead, but even that's not stable now with the tides coming in and flooding the roads. We're living with the sea, but we don't understand it anymore.
>
> —Male fisher, age 63, Ban Khao Thong, Krabi

> We used to catch enough to feed the family and sell in the market. Now the fish are fewer and farther out. We don't have the boats or fuel to chase them. If we lose a net or break an engine during a storm, we might not fish for weeks. We have no insurance, no help. Some of us tried to switch to working in the hotels, but even the hotels are struggling when the sea floods their land. We feel forgotten—like we're just waiting for the next disaster to take more away.
>
> —Female fisher, age 57, Ban Nai Sa, Krabi

In Krabi, tourism businesses are battling direct physical effects on infrastructure and aesthetics, especially those around the beaches of Ao Nang and Railay. In addition to exposing tree roots and causing damage to pathways and hospitality facilities, shoreline erosion has resulted in the loss of beachfront land. Hotel managers and small guesthouse owners interviewed expressed annoyance with the high expense of temporary preventive measures and the recurring financial losses brought on by monsoonal flooding. A yearly ritual, according to several operators, is the seasonal repair of drainage canals and sandbanks. However, these are reactive fixes that provide minimal sustainability or long-term protection.

Furthermore, smaller companies are more at risk than big resorts, including tour guides, long-tail boat operators, and neighbourhood merchants. Tidal encroachment has reduced usable docking space and increased the amount of time lost to weather-related cancellations, according to an Ao Nang boat operator. Even a few missed workdays during the busiest time of year can have a big impact on household income for these unorganized workers. Although some operators have participated in coastal restoration projects, such as establishing natural bamboo barriers or planting sea grapes, these activities are still dispersed and lack adequate funding.

Rice farmers in Nakhon Si Thammarat are supported by the deltaic plains, and their cycles of production are closely linked to soil conditions, rainfall timing, and water availability. Farmers are torn between seasonal flooding and dry-season salinization in flood-prone areas like Pak Phanang and Hua Sai. Field trips showed that farmers are unable to forecast the best times for planting and harvesting rice due to the growing unreliability of traditional calendars. In recent years, delayed monsoons have left crops vulnerable to drought and pest infestations, while early rains have flooded fields, killing seedlings. Interviews with cooperative members show a rising shift toward salt-tolerant rice varieties, but uptake is slow due to higher seed costs and uncertain yields.

The creeping salinization of water and soil is arguably the most harmful, especially in fields close to canals that now experience brackish backflow during high tides. Despite several flushing attempts, a farmer in his 60s in Ban Tha Phaya displayed a soil salt meter reading much over safe standards for rice. His family had converted half of their plot to shrimp farming, but the move had an ecological and economical cost because it called for additional feed infrastructure, ponds, and pumps. Others, caught in cycles of low production and debt, continue farming in increasingly marginal conditions without access to financing or secure land tenure.

A unifying theme throughout these three case studies is the degradation of livelihood security brought on by the instability of natural systems. Every community is reacting, but their reactions are limited due to a lack of institutional support, restricted access to financial resources, and a lack of representation in official planning procedures. For small-scale fishermen, market volatility, navigational hazards, and shifting marine ecology all intersect. Salinization and unpredictable water supply pose a threat to rice farmers' economic independence and generational customs. Environmental deterioration and fragile infrastructure pose a threat to tourist operators' ability to maintain operations and create jobs in the community.

Importantly, these effects are not unique. Those whose lives depend on regular environmental rhythms are particularly vulnerable to these compound vulnerabilities. They are being forced to the limits of their knowledge systems, which have been formed by decades of close contact with the sea, land, and weather. These communities must be viewed as potential co-creators of localized, adaptive methods as hydro-meteorological dangers increase, provided they receive the necessary institutional paths, resources, and recognition.

2.4 Ecological Frontlines: Mangroves, Wetlands, and Natural Buffers

The first line of protection against storm surges, tidal erosion, and rising seas along the Krabi and Nakhon Si Thammarat coastlines is ecological rather than concrete. Coastal woods, wetlands, and mangroves all play an important but frequently overlooked role in protecting communities from hydro-meteorological hazards. By stabilizing sediments, filtering seawater incursion, absorbing wave energy, and offering vital habitat for biodiversity that sustains local livelihoods, these ecosystems serve as protective barriers. However, human encroachment, unsustainable land conversion, and changing climatic baselines (Allam & Cheshmehzangi, 2024) are posing a growing danger to ecological systems in both provinces. This section examines how vulnerability is increased by ecological degradation and how restoration initiatives, frequently spearheaded by local communities, are reclaiming these natural buffers as resilience anchors.

In Nakhon Si Thammarat, vast expanses of deltaic mangrove forests once shielded inland agricultural lands from coastal hazards. A rich estuarine system with mangroves spanning tens of thousands of hectares was found in the Pak Phanang River basin in particular. However, over 60% of these mangroves have been lost or deteriorated during the last three decades as a result of land reclamation, illicit timber harvesting, and conversion for shrimp farming. Inland regions are now vulnerable to daily tidal surges and salt intrusion since many formerly forested tidal zones have turned into bare mudflats, according to field surveys carried out in 2023. Farmers and village leaders in districts like Hua Sai and Chian Yai have repeatedly connected the loss of mangroves, which historically served as sediment stabilizers and salinity filters, to the rising salinization of rice fields.

Natural drainage systems have also been weakened by the loss of wetlands. Freshwater wetlands in the low-lying areas of Nakhon Si Thammarat used to hold onto monsoon runoff and shorten the time when floods flooded. Road building, population growth, and aquaculture infrastructure have drained or fragmented these wetlands, lowering biodiversity and improving water retention. Flooding events become more intense and frequent as these ecological buffers disappear, particularly when storm surges occur in conjunction with periods of high rainfall. Because ecosystem deterioration results in increased exposure and longer recovery times, the province's ability to reduce the risk of disasters is thus closely linked to its ecological health.

Despite these challenges, community-based restoration efforts have emerged as powerful responses. With the help of NGOs and district forestry officials, local organizations in Pak Phanang have started mangrove replantation initiatives utilizing indigenous species including Avicennia alba and Rhizophora mucronata. To determine priority restoration zones, these programs integrate contemporary hydrological mapping with historic biological knowledge. Youth organizations have incorporated planting and nursery management into school curricula in a number of villages, fostering local stewardship and environmental awareness. The protective advantages start within the first two years when root systems stabilize soils and intercept tidal flows, even though regeneration is slow—mangroves take five to ten years to reach full maturity. In addition to being ecological, these bottom-up strategies are sociopolitical since they break cycles of degradation and increase community trust and ability.

Patchy mangrove belts, coastal forests, and coral-fringed islands, which maintain a distinct risk environment, are more prominent in Krabi's ecological narrative. Even while Krabi has not lost as many mangroves as Nakhon Si Thammarat, the city's remaining mangrove areas—especially those along Ban Ao Luk and the Krabi River estuary—are coming under more strain from urbanization, canal dredging, and tourism development. These ecosystems serve as storm buffers and sustain artisanal fishing, but they are frequently left out of tourism master plans or deteriorated by careless waste management and water discharge from neighbouring resorts.

Mangroves are less continuous but just as significant along Krabi's rocky and sandy shorelines. In regions where they are found, mangroves offer vital sediment anchoring that benefits marine life by slowing erosion and lowering turbidity in nearshore waters. In comparison to nearby bare shorelines, mangrove-protected portions of the coastline saw 35–50% less erosion during storm surge events, according to a research study conducted by the Krabi Provincial Office in 2022. However, mangroves are seen as disposable in many of these areas; they are either cleared for construction or filled in to make way for infrastructure and access roads. These choices have long-term consequences since even little mangrove belt deterioration can have a domino effect on the environment and society.

Restoration efforts in Krabi are expanding, albeit unevenly, much like in Nakhon Si Thammarat. To restore deteriorated tidal zones, community forestry organizations have started small-scale replanting and zoning projects, occasionally in collaboration with resort owners and local authorities. In three years, the Department of Marine and Coastal Resources and a local school collaborated to recover around 12 hectares of mangrove in Ban Khao Thong. Participants noted a return of juvenile fish and crabs that sustain local fishing communities in addition to less flooding in surrounding areas. These programs show the promise of nature-based solutions that combine ecological function with local livelihoods, despite the fact that they are frequently dispersed and little funded.

More recently, Krabi has investigated hybrid strategies that combine low-impact engineering with ecological buffers. An integrated model of protection, education, and economic diversification is provided by the use of bamboo fences, floating vegetation mats, and elevated walkways through mangrove zones to both protect and draw eco-tourists. These initiatives have the potential to serve as replicable models for other provinces attempting to strike a balance between development aspirations and ecological preservation.

A recurring theme in both provinces is that robust coastal ecosystems are essential—not optional. The resilience to and recovery from climate risks will depend more on the health of these ecological frontlines. Floodwaters slowdown in areas with a lot of mangroves. Drainage gets better where wetlands are present. Sediment holds where woods grow. The opposite is also true: a type of infrastructure failure is ecological neglect.

Restoring the forests, wetlands, and living shorelines that have protected generations must be the first step in restoring resilience in Thailand's coastal areas. These are living infrastructures of the future rather than artifacts from the past. Communities in Krabi and Nakhon Si Thammarat are acknowledging their role and reweaving the very fabric of coastal resilience, not just restoring ecosystems.

2.5 Navigating Disconnection: Governance Challenges in Coastal Risk Planning

The importance of institutions in anticipating, responding to, and reducing climate-related risks has grown as these hazards spread along Thailand's southern coast. However, there is a widening gap between the realities faced by coastal municipalities and the national frameworks for disaster and climate policy in both Krabi and Nakhon Si Thammarat. This part assesses the institutional framework that oversees risk management in both provinces, looking at the local implementation of planning, zoning, and adaptation policies as well as the areas with the most significant deficiencies.

On paper, Thailand's national frameworks for climate adaption and disaster risk reduction are strong and extensive. In addition to creating its own National Adaptation Plan (NAP) and Disaster Risk Management Master Plan, the nation has embraced important international tools like the Sendai Framework. These documents set high standards for integrated coastal zone management, ecosystem-based adaptation, and community resilience. The implementation of these techniques into workable local governance, however, is still dispersed and uneven, according to research conducted in Krabi and Nakhon Si Thammarat.

Provincial development plans for Krabi recognize how critical it is to manage tidal flooding and coastline erosion, especially in tourist-heavy districts like Ao Nang and Nuea Khlong. However, there is little proof that zoning decisions actually take into account scientific data, such as floodplain mapping, erosion modelling, or salinity risk forecasts. The technical ability to interpret climate projections and access to current geospatial datasets are often lacking in local planning departments. Because of this, building permits for resorts and infrastructure in low-lying, erosion-prone locations are still being granted, frequently without sufficient buffer zones or flood mitigation measures. According to municipal officers interviewed, risk assessments are guided more by past weather trends than by predictive modelling.

The situation is significantly more complicated in Nakhon Si Thammarat. Despite the frequent occurrence of storm surges, tidal floods, and land subsidence in districts like Pak Phanang and Hua Sai, their planning frameworks are still primarily reactive (Curoy et al., 2022b). Provincial disaster offices coordinate emergency responses, but limited integration between environmental, agricultural, and infrastructure departments, overlapping jurisdictions, and irregular budget allocations hinder long-term risk reduction initiatives. For example, there may be minimal cross-departmental collaboration when the Department of Irrigation plans flood control infrastructure, and the Department of Marine and Coastal Resources oversees nearby mangrove restoration projects. Overall resilience is weakened by the conflicting or redundant interventions brought about by this sectoral fragmentation.

The absence of consistent interaction between local governance organizations and scientific institutions is a significant deficiency in both provinces. For Krabi and Nakhon Si Thammarat, universities and research facilities have created thorough vulnerability assessments, hazard maps, and climate impact models, frequently with the help of national or international subsidies. Rarely, though, is this knowledge operationalized. When it comes to accessing or using this data, municipal planning officials point to a lack of funding, training, and continuity. Officers in one Nakhon Si Thammarat district office continued to use anecdotal flood history to inform local preparedness efforts, while a set of printed flood risk maps from a 2017 research study went idle in a cabinet.

There are encouraging institutional innovations in spite of these drawbacks. The Office of Natural Resources and Environmental Policy and Planning (ONEP) collaborated to produce the Krabi Provincial Climate Action Plan, which has started to identify significant coastal risk zones and include ecosystem-based adaptation into local planning principles. The potential for bottom-up planning techniques to supplement top-down frameworks is demonstrated by pilot projects like participatory mapping workshops with Ban Ao Luk communities. However, the scope of these initiatives is still restricted, and they rely significantly on outside assistance.

Since Tropical Storm Pabuk in 2019, Nakhon Si Thammarat has seen a substantial improvement in early warning systems and disaster preparedness exercises. Village chiefs, school administrators, and medical professionals are now regularly participating in seasonal flood predicting meetings, which the local government has started to formalize. However, these efforts continue to focus more on emergency response than on long-term adaptation. Additionally, coastal zoning laws are still out of date and frequently do not take into account the salinity and flood risk conditions of today, let alone anticipated future developments.

The lack of legally binding, climate-informed land use regulations is arguably the most obvious governance divide between the two provinces. Although integrated coastal management and climate resilience (Cheshmehzangi, 2025) are emphasized in national plans, local enforcement mechanisms are inadequate. Environmental zoning regulations are frequently overridden by political pressures from the tourism, aquaculture, and real estate industries. In addition to endangering individuals and property, this regulatory weakening sends a message to communities that immediate financial gain is more important than long-term security.

Improving vertical integration between national and local authorities, strengthening municipal planning units' ability to access, analyze, and apply scientific data, encouraging inter-agency cooperation at the provincial level, and incorporating participatory methods into risk governance are all examples of the institutional changes that will be necessary to close the gap between policy intention and practical implementation. Climate resilience in Krabi and Nakhon Si Thammarat will ultimately need to be co-produced in the cities, villages, and mangrove-lined coasts that deal with the everyday reality of a changing coast, rather than being given purely through policies drafted in Bangkok.

2.6 Reimagining Resilience and Toward Adaptive Coastal Futures: Lessons from Thailand's Shifting Shores

In order to present a set of guidelines for coastal resilience planning in tropical deltaic and tourist-heavy areas, the final section summarizes the research conducted by Krabi and Nakhon Si Thammarat. It lays the groundwork for later chapters on community-based adaptation and planning reform by arguing for a change from reactive disaster responses to proactive, ecosystem-sensitive, and livelihood-inclusive adaptation strategies.

Despite having different locations and means of subsistence, the experiences of Krabi and Nakhon Si Thammarat are similar in that they are both subject to growing climate risks and institutional constraints. From the salt-invaded deltas of Nakhon Si Thammarat to the beaches of Krabi devastated by erosion, this chapter has charted the complicated geography of coastal vulnerability. It has examined how physical geography, meteorological volatility, ecological degradation, and fragmented governance interact to produce a landscape of risk—dynamic, uneven, and deeply human. However, there are important lessons to be learned from this fragility for creating more inclusive, adaptable, and progressive coastal futures.

The most obvious finding is that the current strategy for managing coastal risk is still overly reactive, emphasizing piecemeal protection and emergency response over structural adaptation and long-term planning. It is no longer possible to maintain this reactive posture. Coastal towns now have to deal with the effects of climate change; it is not an occurrence for which they should be prepared. Both provinces show that isolated infrastructural improvements or engineered defences alone only provide short-term respite. Static solutions are rapidly surpassed by dynamic dangers in deltaic ecosystems and along coastlines that are prone to erosion. Instead, a change toward livelihood-centered and ecosystem-sensitive resilience is required, grounded in local uniqueness and the lived experience of those who live in these shifting environments.

Ecological buffers, such as mangroves, wetlands, and tidal forests, are crucial infrastructures for regenerative recovery in addition to serving as first line defences against salinity and tidal surge. Local communities express a revived sense of environmental stewardship and less vulnerability to risks in areas where these systems are intact or being restored. Community-driven mangrove replanting, nature-based erosion management, and hybrid green-grey solutions show that adaptation doesn't have to compromise ecological integrity or local livelihoods in Krabi or Nakhon Si Thammarat. Instead, resilience works best when it is jointly produced by ecosystems, communities, and institutions.

In addition to being economic endeavours, the livelihoods discussed in this chapter—artisanal fishing, rice cultivation, and coastal tourism—also involve socio-ecological interactions. In order to facilitate recuperation and transition, they rely on predictable seasonal cycles, stability in the natural environment, and faith in institutions. However, both institutional negligence and environmental unpredictability are undermining them more. To adapt to shifting tides, fish patterns, and salinity levels, farmers and fishermen are using generational expertise to modify their methods in real time. However, they encounter political and institutional obstacles rather than ecological ones, such as limited access to adaptive funding, a lack of insurance, and no official participation in planning procedures. These are failures of recognition rather than resilience.

As said, governance systems continue to be one of the most important factors in determining how resilient Thailand's coastline will be in the future (Larsen et al., 2011). Despite the existence of national programs and pledges to adapt to climate change, the disparity in local implementation keeps growing. Land use decisions frequently reflect short-term economic incentives rather than long-term risk reduction, municipal planning departments lack adequate funding, and cross-sectoral coordination is irregular. However, there are encouraging indications of institutional experimentation within this fragmented landscape, such as youth-led environmental monitoring, seasonal risk briefings, and participatory flood mapping. These programs need to be expanded as fundamental governance procedures rather than trial ventures.

Resilience is an evolving practice rather than a static endpoint, as demonstrated by the experiences of these two provinces. It needs to be based on both local knowledge and scientific understanding, responsive to community needs, and geographically flexible. Planning needs to include flexibility, scenario thinking, and iterative learning in addition to zoning regulations and danger maps. Above all, it needs to elevate the voices of people who are on the front lines—not just as recipients of protection, but also as creators of their own destinies.

The lessons learned here lay the groundwork for the upcoming chapters. Next, we look at community-based adaptation models, where local actors are redefining what it means to thrive in a changing climate, creating networks, and developing innovative practices instead of waiting for top-down solutions. Krabi and Nakhon Si Thammarat's coastal futures are still being shaped. They are neither reliant on heroism nor destined to fail. The decisions that fishermen, farmers, planners, and students make on a daily basis—decisions that continue to adapt, resist, and rethink their relationship with the sea, the land, and one another—are shaping them.

This chapter ends with an invitation rather than a list of things to do: to view the coastline as a place for negotiation, innovation, and rejuvenation rather than as a line of defence. Futures with adaptive coastlines are achievable, but only if we shift our focus from resilience as survival to resilience as transformation. Thailand may be establishing itself as a leader in climate change in the edge zones where land recedes and tides rise.

References

Aksornkoae, S., & Bird, E. (2010). Thailand: Gulf of Thailand Coast. In *Encyclopedia of the world's coastal landforms* (pp. 1135–1140). Springer.

Allam, Z., & Cheshmehzangi, A. (2024). *Sustainable futures and green new deals* (pp. 1–137). Palgrave Macmillan.

Attavinijtrakarn, T. (2019). *Quantitative microbial risk assessment of water consumption in flood-prone communities, Nakhon Si Thammarat province, Thailand.* https://digital.car.chula.ac.th/chulaetd/8849/

Bidorn, B., Phanomphongphaisarn, N., Rukvichai, C., & Kongsawadworakul, P. (2020). Evolution of mangrove muddy coast in the Western Coast of the Upper Gulf of Thailand over the past six decades. In *Estuaries and Coastal Zones in Times of Global Change: Proceedings of ICEC-2018* (pp. 429–442). Springer.

Cheewinsiriwat, P., Langkulsen, U., Lertwattanamongkol, V., Poompongthai, W., Lambonmung, A., Chamchan, C., Boonmanunt, S., Nakhapakorn, K., & Moses, C. (2024). Assessing coastal vulnerability to climate change: A case study of Nakhon Si Thammarat and Krabi. *Social Sciences, 13*(3), 142.

Cheshmehzangi, A. (2025). Innovation and environmental sustainability: ICT and beach management. In *Urban innovation and sustainability: ICT-based Interventions in Hainan, China* (pp. 59–82). Springer Nature Singapore.

Curoy, J., Ward, R. D., Barlow, J., Moses, C., & Nakhapakorn, K. (2022a). Coastal dynamism in Southern Thailand: An application of the CoastSat toolkit. *PLoS ONE, 17*(8), Article e0272977.

Curoy, J., Ward, R., & Barlow, J. (2022b). *Thai coast project report Work Package 1 Baseline assessment of hydro-meteorological boundary conditions.* https://sites.edgehill.ac.uk/nerc-tcp/files/2022/04/WP1_Baseline-assessment-of-hydro-meteorological-boundary-conditions.pdf

Larsen, R. K., Calgaro, E., & Thomalla, F. (2011). Governing resilience building in Thailand's tourism-dependent coastal communities: Conceptualising stakeholder agency in social–ecological systems. *Global Environmental Change, 21*(2), 481–491.

Scheffers, A., Brill, D., Kelletat, D., Brückner, H., Scheffers, S., & Fox, K. (2012). Holocene sea levels along the Andaman Sea coast of Thailand. *The Holocene, 22*(10), 1169–1180.

Suphawajruksakul, A. (2005). *Coastal Erosion at the Pak Phanang River Basin, Changwat Nakhon Si Thammarat* (Unpublished M. Sc Thesis). Department of Geology, Faculty of Science, Chulalongkorn University.

Chapter 3
Socioeconomic Vulnerabilities and Livelihood Resilience: Community Impacts in Krabi and Nakhon Si Thammarat

3.1 Framing Livelihood Resilience in a Changing Coastal Climate

Given the growing climate risks that disproportionately affect vulnerable coastal regions, livelihood resilience is becoming increasingly important. Livelihood resilience, which is defined as the ability of people, households, and communities to absorb, adapt, and change in response to external shocks, captures how communities manage compound stressors, which are those that result from environmental changes, socioeconomic pressures, and institutional weaknesses all at once. Due to geographic vulnerability, patterns of economic dependency, and ingrained societal vulnerabilities, livelihood resilience and climate risk interact acutely in Southern Thailand, especially in the provinces of Krabi and Nakhon Si Thammarat. To develop an empirically supported understanding of the lived realities of climate-related pressures, this chapter examines these crucial intersections using a rigorous, integrated methodology that combines household surveys, health vulnerability assessments, and mapping of the informal economy.

The eastern coasts of Thailand's Malay Peninsula are home to Krabi and Nakhon Si Thammarat, which serve as prime examples of Southeast Asia's growing climate vulnerability in the face of uneven development. Sea level rise, more intense monsoons, flooding, coastal erosion, and saltwater intrusion are some of the unique but connected climate hazards that these provinces face, and they all exacerbate preexisting socioeconomic vulnerabilities. For example, the tourism-driven economy of Krabi exhibits unique fragility due to seasonal employment cycles, informality, and the vulnerability of workers and tourism infrastructure to coastal hazards (Sriwapee et al., 2025). In contrast, the agriculture-based livelihoods of Nakhon Si Thammarat are still largely dependent on stable weather patterns, which are currently in jeopardy due to ongoing flooding, agricultural land salinization, and declining soil health, all of which are made worse by a lack of government action and disjointed land tenure systems.

A. Cheshmehzangi, *Thailand's Tropical Thresholds*, Urban Sustainability,
https://doi.org/10.1007/978-981-95-4975-7_3

This chapter uses a mixed-methods approach that aims to capture both the quantitative and qualitative aspects of vulnerability and resilience in order to address these issues. First, reliable quantitative data on economic structures, household income variability, migration patterns, and disaster preparedness behaviours are provided by household surveys carried out in representative communities in both provinces. These surveys shed light on the complex patterns of subsistence and market-oriented agriculture in Nakhon Si Thammarat as well as the economic precarity associated with Krabi's reliance on tourists. They record important information such as the seasonality of employment, remittance flows, informal income sources, and households' financial stability in the face of climate-related economic disturbances.

Health vulnerability assessments are a crucial second pillar of the research approach that support household surveys (Tantivess et al., 2019). Because health outcomes rapidly deteriorate in the lack of sufficient housing, sanitation, nutrition, and healthcare access—conditions that are often made worse by climatic impacts—health resilience is closely linked to socioeconomic vulnerability. Stressors to physical and mental health that are common in the populations under study are examined in this assessment, with a focus on informal labourers, older people, and households headed by women. Heatwaves, watery illnesses brought on by flooding, vector-borne infections, and chronic stressors associated with livelihood insecurity provide increased health risks to these populations (Akhtar, 2016). Such vulnerabilities frequently coexist with limited state-sponsored social protection programs and poor access to healthcare facilities in both Krabi and Nakhon Si Thammarat, resulting in even greater socioeconomic marginalization.

The magnitude, scope, and structural features of informality across the study locations are revealed by the third methodological factor, informal economy mapping. In Southern Thailand, informality becomes a major source of income in both rural agricultural settings and urban areas that rely heavily on tourism. Informal workers and businesses are more vulnerable during climate shocks because they usually do not have access to formal insurance, institutional protections, or disaster relief procedures. By highlighting the ways in which informal economies may be able to mobilize social capital and adaptive capacities to mitigate climate-related stresses, mapping these sectors makes it easier to identify economic nodes that are essential for enhancing community resilience.

Recognizing regional similarities and clear local differences is crucial when placing these approaches in the larger Southeast Asian context. Similar climate vulnerabilities, such as rising sea levels, altered monsoon patterns, and more powerful tropical storms, affect coastal communities throughout the ASEAN region (Dong et al., 2024). However, there are significant differences in the historical patterns of land use and urbanization, local economic bases, government structures, and levels of community cohesion. Krabi's strong reliance on tourism, for instance, is comparable to other regional tourist destinations like Bali, Indonesia, and Boracay, the Philippines, where climate-related disturbances also increase precarity in the informal job sector. On the other hand, the agricultural vulnerabilities in Nakhon Si Thammarat are similar to the difficulties encountered by communities in Vietnam's Mekong

Delta, where land-use conflicts and subsistence farming are closely linked to climate change.

In sum, the conceptual framework presented in this section emphasizes how urgent it is to address climate resilience by using sophisticated theories of socioeconomic vulnerability (Cheshmehzangi & Dawodu, 2018). An excellent basis for deciphering intricate dynamics in Krabi and Nakhon Si Thammarat is provided by the integrated research methodology used, which combines household surveys, health vulnerability assessments, and mapping of the informal economy (Cheewinsiriwat et al., 2024). The chapter prepares the reader for a thorough examination of place-specific vulnerabilities and resilience tactics by placing these case studies within larger regional patterns of climate risk and adaptive responses. In the end, this strategy emphasizes that improving livelihood resilience necessitates contextually aware, economically inclusive, and socially equitable approaches, recognizing that resilience-building is essentially about shifting development paths toward sustainability and social justice rather than just reducing short-term risks.

3.2 Tourism, Informality, and Precarity: The Krabi Experience

One of the most well-known coastal destinations in Southern Thailand, Krabi, embodies the dangers and opportunities of economies that rely heavily on tourism. What was once a small fishing village has become a major tourist destination that attracts millions of visitors each year thanks to its globally recognized beaches, islands, and marine ecosystems. However, beneath the surface of wealth is a fragile socioeconomic environment marked by seasonality, unorganized labour markets, and increased susceptibility to disturbances brought on by climate change. The intricate picture of vulnerability painted by Krabi's economic systems, social hierarchies, and environmental fragility is revealed by the research conducted through in-depth household surveys and mapping of the informal economy (see Figs. 3.1, 3.2, 3.3, 3.4, 3.5, and 3.6).

The extremely seasonal nature of Krabi's tourism business is a defining feature of its socioeconomic vulnerability (Calgaro et al., 2009) (Fig. 3.7). Employment in the hospitality, marine recreation, and ancillary service industries is maintained by the peak tourism months of November through April, although economic activity drastically drops during the monsoon seasons of May through October. Residents of Krabi town, Ao Nang, and the nearby islands, such as Koh Phi Phi and Koh Lanta, are heavily dependent on temporary, informal, and part-time work, according to surveys done there. Without official contracts or social security benefits, about 60% of respondents said they worked in informal jobs like food preparation, street vending, boat operations, or freelance guiding. This informality reduces households' ability to maintain steady incomes and act as a buffer against environmental shocks,

Fig. 3.1 New Development at one side and remarkable natural features on the other side of Krabi's coastal areas

Fig. 3.2 Even the most rugged cliffs and rocky edges of Krabi are not immune to the slow but relentless forces of erosion. Over time, wind, rain, and tidal action carve away at their surfaces, reshaping the coastline and revealing the fragility hidden within their imposing forms. This ongoing process serves as a reminder that, in coastal environments, even the hardest stone yields to the persistence of nature

Fig. 3.3 Towering limestone cliffs frame a secluded stretch of Krabi's coastline, their weathered faces telling a geological story shaped over millions of years. Below, emerald waters lap gently against scattered rock outcrops, while a fringe of coconut palms lines the quiet beach. Despite their imposing presence, these cliffs are slowly sculpted by wind, rain, and waves, a reminder that even the grandest coastal formations are subject to the unceasing forces of nature. Note the slow pace of new development is visible in distant

which exacerbates economic insecurity, especially during seasonal downturns and extreme weather events.

These unstable conditions are made worse by climate-related hazards (see Figs. 3.8, 3.9, 3.10, 3.11, 3.12, and 3.13). Shoreline erosion, storm surges, flooding, and unpredictable monsoon patterns are all becoming more common in Krabi, disrupting the island's tourism infrastructure and activities. Both official hospitality workers and unofficial sellers who rely on daily incomes were directly impacted by the significant damage done to local marketplaces, transit lines, and seaside lodgings during recent severe monsoonal floods. According to survey results, almost half of participants reported income losses of more than 50% during these occurrences, which caused many families to face short-term difficulties, accrue debt, and become more food and housing insecure. Informal workers living in coastal areas or flood-prone informal settlements without proper drainage or disaster-resilient infrastructure were especially vulnerable to housing issues.

These housing and economic inequities are exacerbated by health concerns. Significant public health hazards were identified by mapping the informal economy and conducting health evaluations. These risks were made worse by a lack of sanitary facilities, overcrowding in temporary accommodation, and restricted access to

Fig. 3.4 Although relatively limited in scale compared to larger urbanised coasts, tourism-driven and other human activities along Krabi's shores play a significant role in shaping the region's ecological dynamics. The area's unique geography—a network of coastal zones and clusters of islands—means that even small-scale interventions can ripple through interconnected marine and terrestrial ecosystems. From beach landings and boat traffic to shoreline developments and recreational use, these activities influence water quality, habitat stability, and biodiversity, with impacts extending well beyond the immediate coastline

healthcare for migrants and informal workers. Increased rates of waterborne illnesses such dengue fever, diarrhea, and skin infections were noted during floods. Community interviews revealed that mental health costs associated with long-term economic stress, job loss, and unclear futures were prevalent, especially among migrant populations, female-headed households, and younger workers who depended on seasonal incomes.

The results of the poll also revealed inequalities throughout the community, with migrants and ethnic minorities who worked informally being disproportionately vulnerable. These groups often had limited access to state-funded disaster assistance programs, lacked land tenure security, and worked in the most dangerous job niches. Thus, during times of climate stress, informality exacerbates precarity and weakens the resilience of larger communities by intersecting with social marginalization.

In sum, Krabi's economy, which depends heavily on tourism and is characterized by significant informality and seasonal vulnerability, makes the province more vulnerable to climate change. Environmental shocks exacerbate socioeconomic disparities and restrict adaptive capacity by causing health vulnerabilities, housing insecurity, and informal labor markets. Resilience-building initiatives must shift from

Fig. 3.5 New development preparing for the next monsoon season and flooding events that normally cause huge disruption in Krabi and its broader region

Fig. 3.6 Fixing what is left of the previous flood and preparing for the next one in Krabi

Fig. 3.7 The almost-everyday stormy conditions of the monsoon season could make tourism businesses more vulnerable in Krabi

immediate disaster responses to systemic interventions that fortify social protections, formalize employment practices, and advance equitable climate adaptation strategies that put the needs of the most marginalized groups first in order to effectively address these multifaceted vulnerabilities.

3.2.1 Adaptive Strategies in a Tourism-Dependent Economy

Families and communities actively participate in localized adaptive methods to reduce economic and climate risks, despite the vulnerabilities inherent in Krabi's tourism-dependent economy. The results of household surveys and mapping of the

Fig. 3.8 A grounded fishing boat rests on the exposed tidal flats of Krabi, with mangrove forests and towering limestone cliffs in the background. While this scene evokes a sense of calm, it also reflects the vulnerability of coastal livelihoods to flooding (as well as tsunami), storm surges, and shifting tidal patterns intensified by climate change. Low-lying coastal zones, such as this, face compounded risks from sea-level rise, habitat degradation, and the growing frequency of extreme weather events—threatening both the ecological balance and the resilience of local communities. This particularly more visible in coastal areas of islands in Krabi's broader region, only reachable by boats

informal economy highlight a number of crucial behaviours that communities that depend on tourism and informal labour have developed, demonstrating grassroots resilience in the face of ongoing climate uncertainty.

Seasonal migration to urban and peri-urban areas, particularly Phuket, Surat Thani, and Hat Yai, during the off-peak travel months, has been recognized as one important adaptation. Because of their mobility, informal workers—especially men and young adults—can find work in sectors like retail, construction, or services that are less affected by the seasonality of tourism. At least one family member migrates seasonally to supplement incomes, spreading financial risks across geographic locales, according to about 35% of studied households. Even though seasonal migration offers much-needed economic relief, it also brings with it new social problems, like family dissolution and strain on migrants who don't have social protections in their new communities.

Another common adaptive technique among Krabi's unorganized tourism workforce is income diversification. In order to lessen reliance on tourism alone, communities actively pursue a variety of economic endeavours at the same time, combining

Fig. 3.9 Krabi's islands suffer more than just heavy rain and regular storms

Fig. 3.10 Sheltering against storms in coastal areas is a daily thing under this heavy rain in monsoon season in Krabi

Fig. 3.11 Low tide reveals the exposed shoreline of a Krabi island, where boats and transport equipment rest idle against a backdrop of steep limestone cliffs. Such tidal extremes, combined with the threat of stronger storms, sea-level rise, and coastal erosion, highlight the climate hazards facing Krabi's island communities and marine-based livelihoods

small-scale farming, fishing, informal vending, and freelance guiding. According to informal economy mapping, there was a high level of participation in supplementary livelihood pursuits like food processing, handcraft manufacturing, and motorcycle taxi services, which offered extra cash during slow travel seasons or following climate-related disruptions. The significance of promoting economic diversification initiatives within resilience planning frameworks is underscored by the noteworthy findings that households with varied income streams reported higher levels of economic stability, resilience, and faster recovery rates after disasters.

Networks of community-based risk-sharing are essential for enhancing the adaptability of unprotected informal workers. Commonly used social capital methods

Fig. 3.12 On arrival to Rai Leh island, you cannot miss the motionless and silent sight of the old barge and towing boat lying still on the shores. The banner reads "*Sorry for the inconvenience. This barge and towing boat are currently out of service due to damage and are in the process of being repaired. We apologize for the inconvenience and thank you for your understanding*". I feel there is a louder message here beyond just the notes of this banner, something for us to reflect on and think about when the storm hits the shores each time. There is a sense of deliberate loud voice in this uncomfortable silence

include mutual aid networks, communal resource pooling, and informal credit organizations (also known as "Rotating Savings and Credit Associations" or ROSCAs). About 40% of households polled reported regularly participating in these unofficial financial networks, which were crucial for disaster recovery since they quickly provided funds for basic necessities, housing repairs, and medical needs. In order to strengthen social cohesiveness and group resilience, women, who were especially involved in these community networks, frequently organized community assistance programs for childcare, elder care, and disaster preparedness training.

Fig. 3.13 Krabi's coasts are accustomed to the rhythm of daily rains and passing storms, a natural part of life in this tropical environment. Sudden downpours, shifting tides, and short-lived disruptions are woven into the everyday experience, shaping local routines and coastal activities. Despite these frequent interruptions, life along the shore adapts seamlessly—boats adjust their schedules, markets pause and resume, and communities carry on with an enduring resilience that reflects generations of living in harmony with the sea and sky

The increasing popularity of eco-tourism and community-based tourist businesses, which prioritize environmental sustainability and steady revenue generation in order to lessen vulnerability, is a noteworthy adaptive trend. Alternative livelihood models that combine conservation efforts, cultural heritage preservation, and tourism activities less vulnerable to climate-driven disruptions are exemplified by community-led ecotourism initiatives in places like Khlong Thom, Ban Na Teen, and Ao Luk District. These programs, which give workers access to training programs, more stable employment circumstances, and institutionalized community governance systems that fairly divide economic advantages, are frequently supported by local governments, non-governmental organizations, and foreign partners.

These adaptive techniques do have certain drawbacks, though. If destination jobs are still insecure and do not have official protections, seasonal migration runs the risk of sustaining economic instability. Limited availability of reasonably priced credit, market data, and training opportunities limit income diversification strategies. Despite their importance, community-based financial networks might not be able to sufficiently handle structural vulnerabilities at a broader scale or withstand the effects of extreme climate shocks. To scale successfully and fairly, the shift to sustainable tourism models also needs consistent institutional support, sufficient funding, and widespread community involvement.

Overall, the adaptive tactics shown in Krabi represent strong local reactions influenced by social cohesiveness and communal resourcefulness, despite obvious vulnerabilities. Supportive policy measures that codify labour rights, improve access to social security systems, increase livelihood diversification initiatives, and promote fair involvement in tourism development are necessary to strengthen these methods. Resilience frameworks might better enable Krabi's tourism-dependent communities to deal with more uncertain climatic futures by acknowledging and supporting these specific responses.

3.3 Climate, Crops, and Inequality: Rural Exposure in Nakhon Si Thammarat

The province of Nakhon Si Thammarat, which stretches along the Gulf of Thailand, offers a strikingly rural contrast to Krabi's economy, which is centered on tourism. It is mostly an agricultural region, with a strong emphasis on fishing, rice cultivation, fruit orchards, and rubber plantations. Communities here are particularly vulnerable to escalating climate threats like flooding, storm surges, soil salinization, droughts, and coastal erosion because of their rural economic foundation. Household surveys and in-depth agricultural assessments carried out across selected districts, including Pak Phanang, Hua Sai, and Sichon, offer compelling insights into how climatic pressures intersect with existing social inequalities, amplifying vulnerability across rural communities.

Flooding, one of the most frequent climate stressors in Nakhon Si Thammarat, has increasingly devastated agricultural lands and disrupted livelihoods. Because of its location, the province is vulnerable to floods brought on by the monsoon, which are made worse by rising sea levels and the decline of coastal mangrove ecosystems. According to surveys, severe flooding has caused large agricultural losses for about 70% of rural households at least twice in the last five years. Farmers specifically highlighted harm to fruit trees including durian, rambutan, and mangosteen, as well as recurring losses of rice crops. Furthermore, by reducing soil fertility and polluting freshwater resources needed for irrigation, storm surge-induced saltwater intrusion makes agriculture even more vulnerable. Surveys showed that the Pak Phanang Basin, which has historically been one of the province's main agricultural centers, was especially vulnerable, with many households reporting declining productivity as a result of flooding and ongoing soil salinity.

Rural vulnerabilities are exacerbated by sociodemographic factors in addition to these natural constraints. Youth out-migration, aging populations, and a shortage of labor are the main demographic concerns facing Nakhon Si Thammarat. According to household data, a significant percentage of homes are headed by elderly people and frequently depend on remittances from younger family members who work in cities like Bangkok, Surat Thani, or Hat Yai. Remittances were reported by nearly half of rural households questioned, but because urban informal work is so unstable, these financial transfers often fluctuate. The ability of rural households to invest in adaptation strategies like land rehabilitation or flood-resilient agricultural methods is hampered by this heterogeneity.

Insecure land tenure makes rural areas even more vulnerable to the effects of climate change. In Nakhon Si Thammarat, a large number of farming communities own land with no official title or ambiguous tenure status, which restricts their access to government assistance and credit that is essential for implementing sustainable farming methods or enhancing infrastructure. Communities in coastal regions like Hua Sai, where traditional mangrove harvesting, aquaculture, and rice cultivation coexist despite growing ecological stress, are most impacted by this precarity. According to informal economy mapping, a sizable percentage of households surveyed—roughly 35%—were involved in agriculture without a secure land tenure, making them more vulnerable to environmental shocks because they are unable to efficiently access state-sponsored recovery funds or compensation schemes.

The results clearly show how poverty and a lack of state assistance are related. Even though the Thai government has occasionally put in place agricultural compensation plans and flood relief measures, many subsistence-level households still cannot access or use them. Widespread discontent with bureaucratic complications and eligibility requirements that are thought to exclude those who need them most was revealed by local interviews and focus groups. Furthermore, there was a notable lack of or inefficiency in agricultural extension services that could have enhanced community readiness and adaptive practices; these programs frequently benefited wealthy farmers rather than smallholders who were at risk.

Health vulnerabilities are equally important and are made worse by climate-related disasters and a lack of adequate rural healthcare infrastructure. Because stagnant waters provide ideal breeding conditions for disease vectors, rural communities reported frequent epidemics of vector-borne diseases, especially dengue fever, during flood occurrences. Access to medical facilities remains problematic, with outlying villages reporting poor or disturbed treatment during disaster occurrences. Furthermore, chronic health problems, such as arthritis and respiratory disorders, usually stated by older respondents, increase under conditions of excessive humidity and post-flood shape infestations. Household health assessments also revealed a strong correlation between mental health issues and economic instability, loneliness, and ongoing climate changes.

An underlying concern found through household and informal economy surveys is gendered vulnerability, where women disproportionately suffer the brunt of rural hardships compounded by climatic stress. When male family members leave Nakhon Si Thammarat to work, women usually take care of the home, look after dependents, and continue to work in agriculture. As a result, when the environment is disrupted, women's workload and exposure to financial and physical dangers rise dramatically. Female-headed families reported higher levels of food insecurity, limited access to financing, and fewer means to invest in adaptive methods such as flood-resistant building or agricultural diversification.

In sum, rural exposure in Nakhon Si Thammarat captures a complex interaction between long-standing socioeconomic disparities and escalating climate impacts. Existing vulnerabilities stemming from population aging, insecure land tenure, a lack of social protections, and gender imbalances are made worse by climate stressors like flooding, salinization, and reductions in agricultural production. Comprehensive, context-sensitive interventions that incorporate equitable land tenure reforms, climate-resilient agricultural methods, increased social safety programs, and focused investments in rural healthcare facilities are necessary to address these interrelated issues. In rural Nakhon Si Thammarat, cycles of poverty and inequality will persist as climate concerns continue to undermine livelihoods in the absence of such multifaceted responses.

3.3.1 Floating Agriculture, Livelihood Shifts, and Resilience Networks

Rural communities in Nakhon Si Thammarat have progressively used localized adaptive methods to increase livelihood resilience in response to the growing socioeconomic constraints and climate vulnerabilities described above. One noteworthy invention that has gained favour among rural Thai people is floating agriculture, a sustainable farming technique that has historically been used in flood-prone areas like Bangladesh and the Mekong Delta. It is particularly prevalent in districts like Pak Phanang and Hua Sai.

By building raised beds or rafts out of locally accessible materials like bamboo, water hyacinth, coconut coir, and straw, floating agriculture allows crops to flourish even during extended periods of floods. This method has worked especially well for growing vegetables like tomatoes, cucumbers, water spinach, and morning glory, giving households a reliable source of food and extra cash during flood seasons. According to surveys, 20% of households in villages that are extremely vulnerable to flooding reported that they had successfully adopted floating agriculture in the previous five years, citing notable gains in both economic stability and food security. Interviews also emphasized the ways in which local farmers actively teach their neighbours how to build rafts, choose crops, and maintain them, so increasing their resilience and self-sufficiency.

Diversification into aquaculture and integrated farming systems is another important change in livelihood that goes beyond floating agriculture. Households are becoming more involved in shrimp, crab, fish, and mangrove regeneration programs, especially in coastal and estuarine areas. In addition to producing larger profits than conventional rice growing, aquaculture benefits the community by taking advantage of saline or brackish water intrusion. Similarly popular are integrated farming systems that include vegetable gardens, fishponds, and rice cultivation to improve household nutrition, revenue stability, and ecological sustainability. According to household surveys, aquaculture participation has doubled in the last ten years, particularly among younger farmers looking for cost-effective alternatives in the face of diminishing rice harvests.

Adaptive efforts are also supported by resilience networks and community-based cooperative approaches. Informal farmer groups and agricultural cooperatives have developed into vital hubs for resilience, offering forums for group resource sharing, collective bargaining, and disaster assistance. Several effective community-led cooperatives that manage common facilities for rice milling, fertilizer delivery, and storage systems were identified through focus group talks; these cooperatives greatly lower the costs incurred by individual farmers. Furthermore, women-led informal saving and credit clubs have become increasingly visible, providing vital emergency financial safety nets, facilitating quick access to capital for recovery investments, and fostering communal cohesiveness.

But obstacles still exist. Many marginalized households lack the upfront funds, technical know-how, and consistent market access necessary for the adoption of floating agriculture or aquaculture. Similar external institutional support, ongoing financial support, and open governance systems are necessary for cooperative efforts to be sustainable. Therefore, even though Nakhon Si Thammarat's rural communities show a great deal of innovation and resilience-building, effective scaling of these initiatives necessitates focused government policies, inclusive financing sources, and all-encompassing agricultural extension services that address systemic injustices.

In conclusion, community-led initiatives that support climate resilience in rural Nakhon Si Thammarat include livelihood diversification, floating agriculture, and cooperative resilience networks. Although these approaches show the way to equitable adaptation and sustained rural livelihoods, their full potential requires supportive legislative frameworks. For sustainable, inclusive rural resilience in the face of growing climate uncertainties, it is still crucial to acknowledge and support these local adaptive mechanisms through fair resource distribution and improved institutional frameworks.

3.4 Health, Housing, and Social Protection Gaps Across the Coast

Critical health, housing, and social protection deficiencies that continuously increase socioeconomic vulnerability and restrict adaptive capacities along Thailand's southern coastline are revealed by comparing the provinces of Krabi and Nakhon Si Thammarat. Both provinces display convergent patterns of exclusion, vulnerability, and inadequate institutional assistance during climate-induced shocks, despite differences in their main economic activities—informally employed people in Krabi are driven by tourism, whereas those in Nakhon Si Thammarat are dependent on agriculture. This part offers a comprehensive understanding of how housing insecurity, health outcomes, and social protection system deficiencies interact to erode community resilience through the integration of household surveys, mapping of the informal economy, and focused health vulnerability evaluations.

In both provinces, public health vulnerabilities are continuously elevated, primarily due to inadequate healthcare infrastructure and restricted access to medical services, particularly during periods of extreme weather and flooding. Despite Krabi's thriving tourism industry, health services are still dispersed, especially for migrant communities, informal workers, and inhabitants of unofficial coastal towns. According to health evaluations, there are frequent increases in respiratory disorders, waterborne diseases, and vector-borne illnesses during flooding. Because of overpopulation and poor sanitation, informal settlements like Ao Nang and Phi Phi Islands are most affected. Similar to this, rural areas in Nakhon Si Thammarat lack a strong medical infrastructure that may quickly contain disease outbreaks. Widespread flooding in agricultural areas exacerbates chronic health issues, gastrointestinal illnesses, and dengue fever outbreaks among older people. During climate disruptions, households in remote rural districts reported major delays in accessing necessary healthcare, frequently as a result of damaged transportation infrastructure or a shortage of medical personnel.

Another widespread issue in both areas is housing insecurity, which makes people more vulnerable to climatic changes. In Krabi, households with informal employment often live in unstable coastal communities that are susceptible to storm surges, coastal erosion, and flooding. According to mapping of the informal economy, almost 45% of households that rely on income from informal tourism do not have a valid housing tenure, which restricts their access to government-funded relief efforts in times of crisis. Furthermore, migratory populations sometimes live in temporary housing constructions that offer no defence against flooding or storms, increasing the likelihood of relocation and lengthening the recovery period after environmental shocks.

Housing vulnerabilities in rural Nakhon Si Thammarat take diverse forms, but the repercussions are just as dire. In this case, communities' capacity to invest in robust housing structures or take advantage of government housing assistance programs is severely hampered by informal land tenure arrangements. About 50% of rural households live in housing that is susceptible to frequent flooding, with little financial resources or support systems available for relocation or reconstruction, according to survey data from flood-prone districts like Pak Phanang and Hua Sai. Because frequent losses of property and assets significantly impair long-term economic stability, home precarity has a direct effect on households' resilience capacities.

Despite being intended to safeguard vulnerable populations from shocks, social protection services continue to be insufficient, dispersed, and frequently unavailable to marginalized groups. Both provinces point out glaring shortcomings in this regard. Informal workers in Krabi's tourism industry frequently highlighted the difficulty in obtaining state social assistance programs, pointing to the lack of official documents, strict eligibility requirements, and bureaucratic complications. Minority ethnic groups and migrant workers experienced even more marginalization, which exacerbated economic insecurity and increased reliance on unofficial coping mechanisms in times of crisis. Similarly, agricultural households in Nakhon Si Thammarat complained about agricultural compensation and disaster relief programs that were either insufficient, delayed, or inaccessible. These programs often did not reach the most vulnerable farmers because of the strict eligibility requirements linked to formal land titles and asset verification procedures.

Additionally, both provinces regularly emphasized the gendered aspects of vulnerability. Households led by women, who frequently balance caregiving duties with economic pursuits, showed increased susceptibility and encountered major obstacles when trying to obtain official support services and social protections. A significant institutional gap in addressing gender-specific resilience difficulties was highlighted by health vulnerability assessments, which consistently showed higher rates of food insecurity, healthcare hurdles, and economic precarity among households led by women.

Policymakers and stakeholders must give priority to measures that specifically address these interconnected vulnerabilities in social protection, housing, and health in order to effectively increase resilience. It is nevertheless crucial to improve healthcare infrastructure and guarantee fair access to medical care during climate disruptions. Vulnerability could also be significantly decreased by enhancing housing security through formalizing land ownership, making finance for resilient building easily accessible, and guaranteeing inclusive disaster relief procedures. Building resilient coastal communities still requires comprehensive reforms to social protection systems that prioritize inclusivity, streamline eligibility requirements, and specifically reach out to marginalized and informal people.

3.4.1 Institutional Blind Spots and Inclusive Resilience Planning

In particular, institutional blind spots in current frameworks for resilience planning are examined in this subsection along with their consequences for inclusive adaptation tactics. Systemic oversights or inadequacies in governance and social protection systems that unintentionally keep marginalized groups out of resilience-building programs are known as institutional blind spots. Efforts to build inclusive, equitable, and resilient communities throughout Krabi and Nakhon Si Thammarat (Palakhamarn & Kamolvej, 2024) are severely hampered by such oversights.

Even with the implementation of provincial disaster management plans and national climate adaption regulations, women, older people, migrant communities, and informal labourers continue to face discrimination. Respondents to the survey frequently emphasized how the absence of avenues for participation in planning and decision-making undermines the ability of adaptation initiatives to address risks unique to a community or reflect local objectives (Ober & Sakdapolrak, 2020). For example, because of bureaucratic definitions of eligibility associated with official land ownership or recorded work status, resilience efforts often overlooked unstable tenure areas in rural Nakhon Si Thammarat or informal communities in Krabi.

Furthermore, local organizations entrusted with disaster planning and recovery—such as government agencies, agricultural extension services, and medical facilities—frequently lack the tools, expertise, or mandates necessary to proactively meet the many needs of the community. Limited outreach capabilities were revealed by local officials interviewed, especially with reference to informal communities or gender-sensitive programs. As a result, rather than reducing current socioeconomic disparities, resilience programs usually make them worse. This emphasizes the urgent need for institutional changes that put an emphasis on inclusiveness, participatory governance, and fair allocation of adaptation funds.

3.4.2 Discussions and Analysis of Findings

Important insights into the difficulties of climate resilience along Thailand's southern coastline can be gained from a comparative study of socioeconomic vulnerabilities and adaptive methods in Krabi and Nakhon Si Thammarat. The results demonstrate how socioeconomic circumstances greatly influence varying vulnerabilities and adaptive capacities, even though climatic hazards like flooding, coastal erosion, and soil degradation are common to both provinces. In Nakhon Si Thammarat, vulnerabilities are ingrained in agricultural dependency, demographic aging, and insecure land tenure, while in Krabi, they are mainly caused by economic precarity associated with seasonal and informal tourism employment.

Seasonal migration, floating agriculture, livelihood diversification, and community-led resilience networks are a few examples of adaptive strategies that demonstrate notable local resourcefulness and grassroots innovation. However, structural inequality, a lack of social protections, and institutional shortcomings frequently continue to limit these tactics. For example, even though aquaculture and floating agriculture present viable sources of income, institutional support and financial constraints continue to restrict their scalability. In a similar vein, to achieve long-term resilience results, the community-based networks that are essential to disaster recovery in both provinces need more official recognition and assistance.

The cross-cutting theme of gendered vulnerabilities repeatedly surfaced, suggesting that gender disparities must be specifically addressed in climate resilience policies to guarantee that adaptive interventions give women's particular economic, social, and health needs top priority. The significance of participatory governance frameworks that actively include marginalized groups in decision-making processes, promoting inclusive and equitable adaptation strategies, is highlighted by the institutional blind spots found in both provinces.

3.5 Converging Lessons and Diverging Pathways

The comparison of Krabi and Nakhon Si Thammarat in this chapter highlights important lessons learned from their similarities and differences in terms of adaptation pathways in the context of climate resilience in Southern Thailand. Vulnerabilities associated with informality, unstable housing and land tenure, inadequate social safeguards, and a lack of adequate healthcare facilities continuously surfaced as significant barriers to successful resilience building in both provinces. These common vulnerabilities highlight the need for all-encompassing, multifaceted actions that address both the deeper socioeconomic disparities that fuel cycles of vulnerability as well as the immediate climate hazards.

Nevertheless, depending on regional economic systems, population trends, and institutional capabilities, adaptation routes vary greatly. Interventions that prioritize formalizing employment practices, bolstering worker protections, and encouraging diverse, sustainable tourism models that are climate-disruptive are necessary to address Krabi's tourism-driven vulnerabilities (Cheshmehzangi, 2025). On the other hand, the agricultural issues facing Nakhon Si Thammarat necessitate significant investments in climate-resilient farming methods, safe land tenure changes, and improved rural infrastructure that can sustain sustainable lifestyles.

Finally, incorporating localized, livelihood-sensitive adaptation measures into larger institutional frameworks dedicated to social fairness, inclusivity, and participatory governance is necessary to promote true resilience in Southern Thailand's coastal communities. The most promising way to achieve sustainable, resilient, and equitable coastal futures is through integrated approaches that are sensitive to local situations while also considering regional similarities.

References

Akhtar, R. (Ed.). (2016). *Climate change and human health scenario in South and Southeast Asia*. Springer.

Calgaro, E., Pongponrat, K., & Naruchaikusol, S. (2009). *Destination Vulnerability Assessment for Khao Lak, Thailand*. Sustainable Recovery and Resilience Building in the Tsunami Affected Region–SEI Project Report. Stockholm, Stockholm Environment Institute.

Cheewinsiriwat, P., Langkulsen, U., Lertwattanamongkol, V., Poompongthai, W., Lambonmung, A., Chamchan, C., Boonmanunt, S., Nakhapakorn, K., & Moses, C. (2024). Assessing coastal vulnerability to climate change: A case study of Nakhon Si Thammarat and Krabi. *Social Sciences, 13*(3), 142.

Cheshmehzangi, A. (2025). Innovation and environmental sustainability: ICT and beach management. In *Urban innovation and sustainability: ICT-based interventions in Hainan, China* (pp. 59–82). Springer Nature Singapore.

Cheshmehzangi, A., & Dawodu, A. (2018). Sustainable urban development in the age of climate change. In *Sustainable Urban Development in the Age of Climate Change: People: The Cure or Curse* (pp. 157–182). Springer Singapore.

Dong, W. S., Ismailluddin, A., Yun, L. S., Ariffin, E. H., Saengsupavanich, C., Maulud, K. N. A., Ramli, M. Z., Miskon, M. F., Jeofry, M. H., Mohamed, J., & Mohd, F. A. (2024). The impact of climate change on coastal erosion in Southeast Asia and the compelling need to establish robust adaptation strategies. *Heliyon, 10*(4), Article e25609.

Ober, K., & Sakdapolrak, P. (2020). Whose climate change adaptation 'barriers'? Exploring the coloniality of climate change adaptation policy assemblages in Thailand and beyond. *Singapore Journal of Tropical Geography, 41*(1), 86–104.

Palakhamarn, T., & Kamolvej, T. (2024). Revitalizing urban resilience in Thailand: Exploring conceptual frameworks and terminology. *Journal of Architectural/planning Research and Studies (JARS), 21*(2), 263–282.

Sriwapee, K., Kenaphoom, S., Ruangchai, N., Insumran, Y., Pawala, T., Jumroenpat, P., & Kaewlamai, S. (2025). Thailand as a world top tourism destination: Balancing growth with environmental and cultural preservation. *International Journal of Multidisciplinary in Management and Tourism, 9*(1), 113–136.

Tantivess, S., Yothasamut, J., & Saengsri, W. (2019). Utilisation of evidence from Thailand's National Health Examination Survey in policy development: Finding the weakest link. *Health Research Policy and Systems, 17*(1), 104.

Chapter 4
Coastal Urbanization Under Pressure: Climate Threats and Development Dilemmas in Bangkok, Pattaya, and Songkhla

Entry to the Chapter: Living with the Tide of Urban Change
My many visits to Bangkok and its broader region have shown me three very different faces of coastal Thailand, yet each reveal how quickly the tide of urban change can reshape a city's relationship with its shoreline. I walked the edges of forgotten khlongs in Bangkok, which were once the city's lifeblood but were now choked or paved over, their ability to absorb seasonal floods diminished with each concrete encroachment; in Pattaya, you can see the unrelenting march of high-rises and beachside developments pressing against the land's natural contours, funnelling stormwater into narrow streets and leaving little room for nature to absorb the shock of heavy rains; and in Songkhla, which is quieter but no less vulnerable, I stood where mangroves once formed a living shield.

Across these cities, I sensed a shared undercurrent: the tension between rapid economic gains and the slower, less visible work of resilience. While local people acknowledge that heat, erosion, and flooding are natural aspects of the yearly cycle, they also note that these patterns have become more severe and unpredictable. Climate risks don't come here as standalone occurrences; instead, they compound, combining tidal floods with unexpected downpours or the heat of urban canyons with the threat of storm surges. The truth is clear: when governance is disjointed, green spaces disappear, and natural buffers are exchanged for temporary growth, even the best-designed coastal defences suffer. The speed and scope of change are putting these towns to the test like never before, even if they have adapted to survive with heat and water.

During my final trip to Thailand before to finishing this book, I travelled throughout the eastern and southern shores and wide spaces, frequently stopping to take pictures of locations that would not be there for long. These sections of coastline, which were occasionally calm and occasionally clearly crumbling, seemed to be the boundaries between the present and an uncertain future. The earth was already being eroded by the tide in some places; in others, the land appeared brittle, as though it may break up into weaker, isolated areas in a generation. I couldn't help but imagine

A. Cheshmehzangi, *Thailand's Tropical Thresholds*, Urban Sustainability,
https://doi.org/10.1007/978-981-95-4975-7_4

how livelihoods would be uprooted, communities would be forced to relocate inland, and cultural landscapes would be lost to time.

At many of these moments, I stood still, allowing time to weigh on me and wondering what may happen in the ensuing 10, 20, or more years. The trends I saw were not specific to any one place; rather, they mirrored the duplication of coastal urbanization much beyond of Bangkok's immediate vicinity. There are unsettling similarities to this expansion: a steady retreat from ecological preservation, a move toward areas of increased climatic risk, and a continuing on a development trajectory from which there is never a turning back. As a result, the window for effective action is smaller, governance finds it difficult to keep up, and natural buffers are reduced. The narrative of Thailand's future resilience—or vulnerability—will be written here, along these shores.

Each photograph I took on that journey now feels less like a record of place and more like a quiet warning. The silhouettes of half-finished developments, the reclaimed plots awaiting construction, the stripped mangrove edges—all of them seemed to speak of a present racing ahead without pausing to measure the cost. In Bangkok and its broader and very much extended region, the skyline stretches further into its floodplains, trading natural storage basins for concrete expanses. Standing in these spaces, I felt the dissonance between the promise of growth and the fragility it leaves behind. In two decades, perhaps, these will be stories of relocation of neighbourhoods moving uphill, of livelihoods redefined, of landscapes altered beyond recognition.

This is the reproduction of a model that undermines its own foundations, not just the expansion of urbanization. Every action without ecological awareness contributes to the climate crisis and creates a vicious circle of dangers that are increasingly difficult and costly to reverse. The idea that the course is not totally set, however, is also there in these peaceful moments by the sea. If we decide to rethink how these coasts develop, the images captured today may continue to serve as reminders of what was preserved rather than as monuments.

4.1 Urbanization and Exposure: Setting the Coastal Context

Over the past few decades, Thailand's Gulf of Thailand coastline has seen tremendous urban transitions marked by substantial land-use shifts, economic diversification, and fast population expansion (Tontisirin & Anantsuksomsri, 2021). The extent and complexity of exposure to climate-related risks have increased due to coastal urbanization, which is mostly concentrated around important metropolitan and regional centers like Bangkok, Pattaya, and Songkhla. These cities' urban population expansion reflects larger regional patterns, such as a consistent rural–urban migration fuelled by increased infrastructure, tourism, industrialization, and economic opportunities. In the past, traditional livelihoods like fishing, farming, and maritime trade had an impact on the gradual development of cities (Wang et al., 2022). However, the urban landscape has seen significant change since the late twentieth century due to increased urban expansion, intensified infrastructure, and commercial and industrial land uses.

In terms of population, the area that includes Bangkok, Pattaya, and Songkhla is home to more over 15 million people, or almost 22% of Thailand's total population (Huang, 2019). With over 10 million people living in its metropolitan area, Bangkok alone puts strain on the city's limited space and natural resources. In contrast, Pattaya's population grew rapidly as a result of tourism-driven development, transforming it from a small coastal village to an urbanized resort city that welcomes over 2 million domestic and foreign tourists each year. With a metropolitan population of about 500,000, Songkhla, the smallest of the three cities studied, has maintained relatively moderate urban growth (Sahacharoenwat, 2024), mainly due to regional trade, fisheries, and its advantageous location close to the Malaysian border.

These cities exhibit a variety of urban morphological patterns in space. Bangkok's urban growth has traditionally followed patterns of water-based settlement along the delta of the Chao Phraya River, where the city's urban fabric was traditionally organized by khlongs, or canals (Jongkroy & Thongbai, 2019). Significant peri-urban sprawl, which is defined by dispersed residential and industrial areas that stretch eastward and encroach on natural floodplains and former agricultural land, is now a feature of the city's spatial shape. In addition to a large number of high-rise developments along the coast and an unofficial urban periphery that extends into nearby hillsides, Pattaya has a linear spatial form that runs north–south along its coastal strip. The urban form of Songkhla is heterogeneous, with a historic core centered on the port and fishing communities, as well as dispersed suburban growth and industrial districts encroaching on delicate wetland ecosystems close to Songkhla Lake.

Every city has unique economic factors influencing the paths of urbanization. Thailand's main economic hub, Bangkok, is greatly impacted by the manufacturing, financial, commerce, logistics, and service industries. The city is home to important industrial areas like Lat Krabang and Bangpoo, which are crucial to the country's economic success but are still very susceptible to flooding and subsidence (Yokota et al., 2020). The tourist, hospitality, real estate, and entertainment sectors are the main drivers of Pattaya's economy. Rapid urbanization brought on by tourism has exacerbated environmental degradation, land-use conflicts, and infrastructure strains. Agriculture, logistics, fisheries, and maritime trade all contribute to Songkhla's economy, which is of regional importance. Urban-industrial land conversion and settlement growth near coastal lagoons and wetlands have been fuelled by the expansion of port infrastructure, fishing operations, and seafood processing.

Increased biological fragmentation, fewer green and blue areas, and more impervious surfaces are all results of land-use change in these coastal communities. Mangrove forests, wetlands, and natural drainage areas have significantly decreased, according to satellite-based analysis, especially in the vicinity of Songkhla's lagoon-side urban edges, Pattaya's hillside developments, and Bangkok's eastern periphery. The loss of permeable land in Bangkok makes heavy rainstorm events worse and significantly raises the risk of flooding. Similar to this, uncontrolled slope developments in Pattaya limit the amount of rainfall that can infiltrate, which frequently results in flash flooding. Due to Songkhla's urban growth into wetland areas, ecological buffers that are necessary for maintaining water quality and preventing storm surges have been jeopardized, making the city more vulnerable to coastal erosion and sea level rise.

Thus, there is a close connection between urbanization trends and climatic vulnerability. Urbanization along the coast immediately lowers ecological resilience and raises vulnerability to climate risks. In addition to diminishing ecosystem services, land reclamation, natural wetland drainage, and permeable surface reduction increase the likelihood of flooding, storm surges, and heat extremes. Furthermore, the concentration of people in locations that are vulnerable to hazards intensifies the possible effects on people, the economy, and infrastructure, posing serious adaptive problems for policymakers and urban planners.

Furthermore, climate resilience has rarely been incorporated into the design stages of infrastructure development throughout these cities. The pace of urbanization outpaced the proactive risk management capabilities of city planning and governance systems. Infrastructure vulnerabilities and socioeconomic disruption are caused by reactive rather than proactive management brought on by this governance gap. Long-term vulnerabilities have been exacerbated by investments that have often placed a higher priority on urban expansion and economic growth than on sustainable urban design principles.

Thus, placing Bangkok, Pattaya, and Songkhla's coastal urbanization within Thailand's larger socioeconomic and historical contexts enables a more nuanced understanding of the ways in which urban growth patterns interact with climatic hazards (Palakhamarn & Kamolvej, 2024). The comprehensive methodology used here highlights how land-use dynamics, infrastructure decisions, urban governance, and demographic trends all critically interact to influence these coastal communities' resilience and exposure. Understanding these relationships is essential for directing transformative initiatives toward sustainable and climate-adaptive urban futures as Thailand continues its current trajectory of coastal development.

4.1.1 Coastal Growth Trajectories: Comparing Bangkok, Pattaya, and Songkhla

Significant urbanization has increased the vulnerability of infrastructure and people to climate hazards in Thailand's Gulf coast cities. Urban environments have been drastically altered by urban expansion brought on by demographic changes, economic diversification, and infrastructure advancements, especially in Bangkok, Pattaya, and Songkhla. These cities' urban population increase is consistent with larger rural-to-urban migration patterns driven by economic opportunities in trade, tourism, and industry. At first, traditional fishing, farming, and maritime trade had a significant impact on the slow urbanization of coastal Thailand. However, during the second half of the twentieth century, urban structures have been drastically altered and vulnerability to the effects of climate change has increased due to enhanced commercial and industrial land uses (Cheshmehzangi, 2023) and accelerated urban sprawl.

With more than 10 million residents, the Bangkok Metropolitan Region (BMR) is the most populated urban area in Thailand. With over two million tourists annually, Pattaya's population has grown significantly due to its tourism-driven development, while Songkhla, with its strategically located port and fishery, maintains modest growth and is home to about 500,000 people, making it a crucial regional economic center. A comparative perspective of Thailand's coastal urbanization is provided by the variety of urban growth processes found in these cities.

Traditional canal networks (khlongs) and vast peri-urban development into eastern floodplains have been hallmarks of Bangkok's coastal growth trajectory. Traditionally built around water-based livelihoods, industrial parks and suburban housing are becoming more and more important in today's urban growth, frequently at the expense of ecological and agricultural land. The previous 20 years have seen a significant expansion into flood-prone areas, according to GIS data, increasing the risk of chronic flooding and subsidence for infrastructure. Over 70% of wetlands have been lost since the late twentieth century due to urbanization, which has drastically decreased natural flood retention areas.

The tourism and real estate sectors have a major impact on Pattaya's urban growth, propelling the city's dense and quick coastline development. Informal developments are being forced onto neighboring slopes by the intense construction of resorts, condominiums, and commercial establishments along the straight beach strip. Spatial data show significant increases in impervious surfaces and declines in wooded areas, which directly exacerbate flash floods and diminish natural stormwater management capabilities. Built-up areas have increased by about 60% since 2000, according to urban land use maps.

The mild urbanization of Songkhla is mostly a result of economic activity related to agricultural, fishing, and maritime trade. Here, urbanization frequently overflows into delicate ecological zones like Songkhla Lake and nearby wetlands, leading to serious ecological damage. Satellite image analysis shows distinct patterns of encroachment into mangrove forests and wetlands, which hinders their capacity to successfully reduce coastal erosion and storm surges. Over the past few decades, around 40% of the original marsh lands surrounding Songkhla Lake have been urbanized.

4.2 Climate Threats in the Urban Coastal Zone

The overlapping climatic threats that the three cities face are evaluated in this section. It draws attention to the growing urban heat islands brought on by less greenery, the escalation of urban flooding brought on by sea level rise and heavy rainfall, and the long-term risk of land subsidence brought on by groundwater extraction and soil compaction.

Bangkok faces chronic flooding and severe subsidence largely resulting from excessive groundwater extraction to sustain urban and industrial demands. Geospatial subsidence monitoring and historical groundwater extraction data have revealed sinking rates of up to 2 cm annually in some locations, seriously jeopardizing flood control infrastructure. The high vulnerability of urban infrastructure is demonstrated by the increasing disruption of transportation networks, industrial zones, and residential neighbourhoods caused by annual flooding events. At least 18 significant flood episodes occurred in Bangkok between 2011 and 2020; several of these events occurred during periods of high tide and overloaded the city's inadequate drainage infrastructure. The threshold for disaster has been further lowered by the combination of aged drainage, decreased surface permeability, and sea-level rise. The Bangkok Metropolitan Administration (BMA) reports that more than 40% of the city is below two meters above sea level, making it more vulnerable to flooding from both rivers and rivers. Figures 4.1, 4.2, 4.3, 4.4, and 4.5 also show how the city's southern edges are getting closer to the coastal areas; a sort of coastal urbanization stretched outside the city and beyond.

Bangkok has also seen an increase in the effects of urban heat islands (UHIs), with Ratchathewi and Din Daeng—areas with high construction densities and little vegetation—showing the most noticeable heat differentials. Bangkok's urban districts

Fig. 4.1 Bangkok's southern urban and outskirt regions towards the Gulf of Thailand in 1985 (*Source* Map generated by the author using Google Earth Pro, from July 2025 maps of the region—openly available by 2025 Maxar Technologies and 2025 Airbus)

Fig. 4.2 Bangkok's southern urban and outskirt regions towards the Gulf of Thailand in 1995 (*Source* Map generated by the author using Google Earth Pro, from July 2025 maps of the region—openly available by 2025 Maxar Technologies and 2025 Airbus)

Fig. 4.3 Bangkok's gradual urban growth towards the southern coasts in 2005 (*Source* Map generated by the author using Google Earth Pro, from July 2025 maps of the region—openly available by 2025 Maxar Technologies and 2025 Airbus)

Fig. 4.4 Bangkok's faster urban growth towards the southern coasts in 2015 (*Source* Map generated by the author using Google Earth Pro, from July 2025 maps of the region—openly available by 2025 Maxar Technologies and 2025 Airbus)

Fig. 4.5 As it stands, Bangkok's urbanized region has already reached the coast, as recorded in 2025 maps (*Source* Map generated by the author using Google Earth Pro, from July 2025 maps of the region—openly available by 2025 Maxar Technologies and 2025 Airbus)

are often 4–6 °C hotter than the surrounding green areas, according to satellite-derived land surface temperature (LST) data. The elderly and low-income populations living in unofficial housing with inadequate ventilation have been disproportionately affected by these temperature extremes, which have also increased the energy demands for cooling.

Widespread impervious surfaces and uncontrolled hillside development contribute to Pattaya's frequent flash floods. Particularly during monsoon seasons, hydrological modelling indicates lower infiltration capacities and higher runoff coefficients. In high-density areas like South Pattaya, the lack of vegetative buffer zones further restricts absorption during periods of intense rainfall, converting streets into makeshift rivers. According to infrastructure surveys, there are insufficient retention basins or secondary drainage channels in a number of catchment zones. Flood velocity and inundation depth are made worse by the city's topography, especially the height difference between the coastal plains and the eastern hills.

Additionally, the consequences of urban heat islands have gotten worse, raising ambient temperatures considerably and putting a load on public health systems. The average temperature has increased by about 1.5 °C during the past 30 years, according to recent climate data. In central Pattaya, where impervious coverage surpasses 80%, the surge is more pronounced. There have been more reports of heat-related ailments as a result of heat stress; local clinics have noted seasonal increases in respiratory problems, heatstroke, and dehydration. Additionally, relying too much on air conditioning leads to a vicious cycle of higher emissions and energy use.

Low-lying coastal communities in Songkhla are seriously threatened by sea level rise, as erosion along the coast has accelerated in recent decades. Significant shoreline retreat is shown by data from coastal monitoring systems, which is exacerbated by the loss of mangrove and wetland buffers, which historically lessened the effects of erosion and storm surge. Residential areas and coastal infrastructure are at risk due to the average annual retreat of the shoreline, which is around 3 m. Coastal erosion has been particularly severe in Muang District and areas adjacent to Songkhla Lake, where protective natural barriers have been fragmented by development.

Furthermore, Songkhla's urban drainage system is still in its infancy in comparison to the city's growing area, which causes localized flooding during seasonal downpours. Particularly vulnerable are low-lying informal communities close to Songkhla Lake. The Thailand Meteorological Department's climate estimates indicate that extreme rainfall events would become more frequent and intense, with possible rainfall peaks increasing by as much as 20% by 2050. Such intensification could lead to long-term waterlogging, infrastructure failure, and the eviction of vulnerable communities if immediate adaptive measures are not taken.

Together, these three cities represent distinct yet similar paths of climate exposure, influenced by their respective ecological contexts, patterns of development, and infrastructure reactions. In Thailand's coastal cities, the compound nature of risks—heat, flood, erosion, and subsidence—requires an integrated and proactive approach to resilience planning.

4.3 Infrastructure Under Strain: Assessing the Adaptive Gap

Bangkok, Pattaya, and Songkhla's infrastructure has been put to the test by the city's rapid development. Systems for coastal defence, green infrastructure, and drainage are frequently out-of-date or underdeveloped, falling behind the growing demands of cities and climate change. Extreme weather events make this infrastructure deficiency evident, as cities frequently experience flooding, infrastructure breakdowns, and economic disruptions. Critical gaps in resilience planning, such as poor maintenance and a lack of integration of green infrastructure solutions, are highlighted by infrastructure assessments.

In Bangkok, the city's historical drainage infrastructure—based on an elaborate canal (khlong) system—has become increasingly dysfunctional due to both physical blockage and policy neglect. These canals, which were once the main thoroughfares of Bangkok's water-based urbanism, are now blocked by road extensions, obstructed by unauthorized buildings, and packed with trash. Although the system's effective conveyance capacity has been much diminished, it is still potentially expansive with more than 700 km of open canals. Nearly 30% of the secondary and tertiary drainage canals in the city run at less-than-ideal flow rates, according to the Bangkok Metropolitan Administration (BMA). The vulnerability of this system was demonstrated by flood disasters that occurred during the 2011 and 2022 monsoon seasons, when tidal surges and combined rains caused numerous areas to be submerged for days. Furthermore, Bangkok's water gates and drainage pumps are dispersed unevenly and frequently overloaded, especially in low-lying neighbourhoods like Bang Khen and Lat Krabang.

Despite the emergence of some creative green infrastructure projects, like as the Chulalongkorn University Centenary Park, which serves as a flood retention landscape, these initiatives are still isolated and purely symbolic. Neither the requirement for large-scale hydrological system integration nor the larger infrastructure deficiency are addressed by them. According to experts, Bangkok would have to replace more than 1200 km of subterranean drainage pipelines and build an extra 15 million cubic meters of retention capacity to be sufficiently equipped for the future precipitation patterns predicted under IPCC RCP 8.5 scenarios.

The infrastructure of Pattaya is confronted with distinct yet equally urgent difficulties. The topography of the city, which is built along a slender strip between hills and the seashore, directs runoff toward the city center. However, stormwater surges during the rainy season cannot be controlled by the current drainage system, which consists of shallow channels and culverts with inadequate dimensions. Flash floods have become more frequent, frequently occurring just minutes after heavy rains. In June 2023, a 90 mm downpour in less than two hours flooded more than 60% of central Pattaya. The primary storm drains' sedimentation and poor upkeep were attributed by the local authorities. Illegal hillside construction has also made runoff worse, increasing the amount and velocity of water entering the city's low-lying areas.

These problems are exacerbated by Pattaya's lack of green area. Its absence of natural absorption zones is seen in the city's 3.2 square meters of public green space per person. High-value real estate and tourism infrastructure have taken precedence over sponge city concepts in development patterns. Real estate developers' lobbying caused a proposed municipal plan for hillside buffer zoning and urban wetland restoration to fail in 2021. Despite the introduction of a few retention ponds on the city's eastern side, its usefulness during periods of high floods is undermined by their improper connection to the drainage grid as a whole.

In contrast, Songkhla's infrastructure vulnerabilities are directly linked to its ecological setting and coastline exposure, despite its lesser size. Due to its proximity to Songkhla Lake and the Gulf of Thailand, a large portion of the city is naturally susceptible to storm surges and tidal flooding. However, no comprehensive plan for coastal protection exists. The beachfront is dominated by ad hoc concrete revetments and sea barriers, although their structural soundness is deteriorating. Waves have been known to overtop during periodic monsoons, causing damage to a number of areas close to Old Town and Hua Khao beach. Additionally, by interfering with the movement of silt, these hard infrastructures exacerbate erosion downstream in exposed places.

Songkhla's drainage system is disjointed, especially in the unofficial communities close to the lake. Many places rely on open ditches or direct discharge into natural water bodies because they lack even the most basic stormwater control features. These activities endanger fisheries and aquatic biodiversity in addition to increasing the risk of floods and deteriorating Songkhla Lake's water quality. Some districts have community-driven maintenance programs, but they lack funding and frequently receive no assistance from municipal technical services.

One of the main obstacles to infrastructure resilience in the three cities is governance. The Royal Irrigation Department, the Ministry of Interior, and the BMA all have overlapping authorities in Bangkok, which leads to regulatory misunderstandings and poor coordination. Due to Pattaya's semi-autonomous status within Chonburi Province, environmental regulations are not strictly enforced, and it is challenging to organize coordinated responses. Underfunded infrastructure planning bureaus and a lack of technical skills at the municipal level plague Songkhla.

Importantly, climate adaption scenarios are rarely incorporated into the current infrastructure development in any of the three cities. With little consideration for future hydrological and thermal changes, the majority of urban infrastructure masterplans were created using historical climate baselines. Long-term planning is also hampered by financial limitations. Transportation and tourism amenities receive a large portion of public infrastructure spending, while green infrastructure, climate-resilient water management, and coastal protection continue to receive insufficient funding.

More than just technological advancements are needed to address these issues. It calls for a change to cross-sectoral, systemic infrastructure governance that emphasizes resilience and appreciates ecological services. Planning precision and responsiveness can be increased by combining information from community-based risk mapping, urban hydrology, and remote sensing. Furthermore, operationalizing infrastructure that is both functional and adaptable requires enhancing institutional capacity at the local level through community involvement, interagency collaboration, and training.

4.3.1 The Limits of Engineered Resilience: Drainage, Green Cover, and Coastal Defences

Urbanization and infrastructural failure have seriously jeopardized Bangkok's long-standing reliance on khlongs for drainage. The scope and efficacy of initiatives to implement contemporary drainage systems and urban green areas, such city parks and retention ponds, have been constrained. Recent flood event analysis highlights enduring shortcomings and highlights the pressing need for systemic improvements.

The volume and velocity of surface runoff have increased due to the substantial reduction of pervious ground cover caused by the fast growth of built-up surfaces, which now make up more than 80% of the land use in central Bangkok. Bangkok has made strides in incorporating smart flood forecasting systems; nevertheless, these tools are still limited in scope and not fully integrated into a drainage scheme. Notably, large-scale infrastructure initiatives like the Rama IX Flood Tunnel have reduced flooding in some locations; nevertheless, these costly projects are dispersed unevenly and move slowly. Swales, green roofs, and vegetated drainage easements are examples of smaller-scale retention and infiltration-based solutions that are frequently viewed as decorative accents rather than necessary urban infrastructure.

Bangkok's vulnerability is further increased by the separation of land-use policy from khlong restoration. Even though some districts have implemented khlong dredging and rehabilitation initiatives, continued canal-side building and urban expansion continue to restrict hydrological function. In high-risk neighbourhoods like Bang Sue and Lat Phrao, more than 500 unlawful buildings block khlongs, lowering their channel effectiveness by almost 40%, according to flood control authorities. The need to combine grey and green infrastructure measures is emphasized by the discrepancy between engineered drainage capacity and real runoff loads, particularly during seasonal peaks.

During the busiest travel peak seasons, Pattaya experiences frequent drainage overloads, which are made worse by a lack of green space and extensive hillside development. Flood frequency data and infrastructure performance studies show that the majority of the current infrastructure is based on engineering sea walls and drainage pipelines, which often break during periods of heavy rainfall.

The lack of slope stabilization measures in new hillside developments and careless construction methods exacerbate the issue even more. Drainage outfalls in communities like Pratumnak Hill are frequently too small or out of alignment with the natural contours, which causes erosion and damages infrastructure during periods of intense precipitation. In upstream areas, the city has tried channelization and concrete check dams, but these measures are short-term and only provide a small amount of absorption or delay. Over 70% of runoff from upland developments enters urban cores directly, without being intercepted or retained, according to municipal data.

The majority of green infrastructure projects in Pattaya lack proper maintenance and hydrological modelling, making them a peripheral aspect of the city's urban design. Due to inadequate enforcement, the 2022 trial project on Third Road, which sought to create a linear park and bio-retention trench, has seen encroachment and sedimentation. In addition to causing obvious coastal recession over the past ten years, Pattaya's reliance on hard infrastructure, such as sea walls along Wong Amat Beach, has also resulted in the scouring of nearby, unprotected areas. Long-term geomorphological changes have been warned about by environmental impact studies, but institutional support for alternatives like dune restoration or mangrove replanting has been weak.

Ecological deterioration and poor upkeep plague Songkhla's coastal fortifications. The city's natural flood and erosion mitigation capabilities are diminished by the loss of mangroves along shoreline areas and the lack of inland green infrastructure. The functioning of the coastal buffer has significantly declined, according to ecological assessments, which calls for integrated ecological-infrastructure interventions.

The use of inexpensive, concrete-based interventions has not been enough to mitigate dynamic coastal processes in Songkhla. An estimated 35% of mangroves have been lost over the last 25 years, leaving the coastline vulnerable to saline intrusion and wave action, which has deteriorated both natural and man-made resources. Areas near Songkhla Lake's southern side are increasingly affected by flooding during monsoonal high tides, especially where informal housing encroaches on once wetland territory. The infrastructure in this area is both financially and functionally stressed; drainage canals are frequently manually cleared, lack engineering, and are not connected to any official stormwater grid.

Songkhla presents unrealized potential for integrated coastal zone management in spite of these drawbacks. In high-risk inland communities, research partnerships with Prince of Songkla University have investigated hybrid flood protection methods that combine porous paving, restored mangrove corridors, and soft embankments. However, scale-up has been slow due to limited budgetary commitments and unclear jurisdictional mandates.

When systems are put to the test by compound climate shocks, like consecutive heat waves and flooding or tidal surges during rainfall events, the limitations of merely built resilience are demonstrated in all three cities. Infrastructure that disregards long-term maintenance requirements, community co-benefits, or ecological feedbacks is becoming more and more unsustainable (Allam & Cheshmehzangi, 2024). Mangrove regeneration, urban forests, rain gardens, and blue-green corridors are examples of nature-based solutions that must be integrated in order to adapt to the current climatic trajectory.

4.4 Land Use Conflict and Informal Urbanism

This section looks at the spatial tensions brought about by uncontrolled land use, rapid development, and the expansion of informal settlements, especially in regions that are hazardous and environmentally sensitive. It draws attention to the ways in which urban inequality, displacement, and these forces interact. In addition to physical growth, Thailand's coastal cities' rapid urbanization has resulted in disputed and frequently chaotic spatial arrangements, especially in areas that are hazardous and environmentally sensitive. Significant land-use conflicts have been brought on by the spread of uncontrolled development and the gradual expansion of informal settlements in Bangkok, Pattaya, and Songkhla. Deeper socioeconomic divides, speculative real estate activities, and governance shortcomings in zoning, enforcement, and land distribution are all reflected in these conflicts.

Bangkok's landscape has been transformed into disjointed mosaics of formal and informal land usage as a result of urban growth gradually consuming floodplain buffers and outlying agricultural areas. Alongside unplanned communities where migrants and the urban poor look for affordable accommodation, high-value buildings like shopping malls, logistics centers, and gated residential complexes have appeared. Land conversion has accelerated in Bangkok's eastern outskirts, including neighbourhoods like Nong Chok and Min Buri, without being logically incorporated into the city's flood management strategy. Low-income populations are more vulnerable because they frequently relocate to flood-prone areas with inadequate infrastructure after being uprooted from inner-city neighbourhoods.

The dynamic in Pattaya is different. Speculative development has favoured luxury apartments, hotels, and commercial plazas, frequently at the price of environmental integrity, driven by tourists and foreign investment. In addition to obstructing natural drainage channels, hillside encroachments—especially in the vicinity of Pratumnak and the northern slopes—have caused erosion and slope instability (Protong, 2016). In marginal areas like canal banks, utility easements, and abandoned lots, informal housing—often inhabited by migrant workers in the construction and service sectors—has spread. In addition to lacking essential utilities, these places are becoming more vulnerable to heat stress and flooding.

Land-use conflicts are more strongly linked to Songkhla's natural resources, particularly the mangrove and wetland areas that round Songkhla Lake (Anukularmphai et al., 2005; Tabucanon, 2004). The growth of industrial estates and uncontrolled habitation have increased urban overflow into these areas. On formerly wetland plots, informal housing clusters have grown, upsetting hydrological flows and destroying important ecosystems. Low-income inhabitants are forced into these ecologically vulnerable areas as the formal land supply becomes limited, escalating ecological degradation and disaster vulnerability. The issue has been made worse by outdated land-use planning maps and poor coordination between environmental and municipal authorities.

These geographical tensions are a reflection of deeper political and economic institutions that put short-term growth ahead of long-term stability; they are not just technical errors. Although it is frequently presented as chaotic, informal urbanism is actually a survival tactic used by underprivileged groups shut out of official housing markets. However, its establishment in high-risk areas without sufficient planning assistance jeopardizes ecological sustainability and human security.

4.4.1 Encroachment, Real Estate Pressures, and the Displacement of Ecologies

Some of the most pressing issues facing urban development in Thailand's coastal cities are defined by the interaction of ecological displacement, informal land occupancy, and real estate speculation. These processes are physically and politically rooted in urban growth patterns that put economic acceleration ahead of social or ecological integrity in Bangkok, Pattaya, and Songkhla. The mechanisms and effects of these pressures are examined in this section, with special attention to how they force vulnerable populations into environmentally dangerous areas and cause delicate ecosystems to degrade irreparably.

In Bangkok, land encroachment has become a defining feature of peri-urban expansion, particularly in low-lying areas such as Lat Krabang, Nong Khaem, and eastern Thawi Watthana (Menakanit et al., 2022). Once characterized by rice paddies, canals, and agricultural wetlands, these districts have experienced significant change as a result of warehousing, logistics, and speculative residential construction. Water retention zones that were once flood buffers have been broken up by the development of commercial estates and gated housing enclaves on agricultural land. The goals of regional planning are undermined when private developers routinely negotiate ad hoc land-use licenses with municipal authorities in the absence of a coordinated urban expansion boundary or enforced zoning rules. Low-income migrants migrate on marginal areas along khlongs or railway lines as a result of these measures, frequently without tenure security or access to infrastructure, when prime land becomes unaffordable due to its privatization.

Although they provide access to urban livelihoods, these informal settlements increase Bangkok's vulnerability to climate and hydrological hazards. Dense, uncontrolled housing constructed on reclaimed canal edges or along embankments is home to communities like Bang Bua and Khlong Toei, which impede water flow and increase the risk of flooding during monsoon peaks. Serious legal issues have also been brought up by the encroachment on public waterways, especially when state agencies try to clear informal settlements in order to restore canals or upgrade infrastructure. Evictions that follow, frequently carried out under emergency orders, uproot already vulnerable groups and drive them farther into the environmental fringe.

The trajectory in Pattaya is different but just as dangerous. High-rise hotels and condominiums now dominate the skyline in this rapidly verticalizing city, which is one of the tourist destinations in Southeast Asia with the quickest rate of growth.

These developments have replaced traditional lowland ecosystems, forested ridgelines, and natural drainage routes, even if they also hold the promise of economic vigour and international investment. The development of hillside areas close to Pratumnak and Naklua has changed the stability of the slope and destroyed vegetation that is essential for controlling surface runoff and regulating the microclimate. Projects have frequently gotten around environmental impact assessments or gotten past gaps in the local permitting procedure.

At the same time, Pattaya's informal dwelling market keeps growing on abandoned or unused land parcels, like those beneath power lines, next to building sites, or in seasonal drainage channels. Generally speaking, these places are unsuitable for human habitation because they frequently experience seasonal flooding, land subsidence, or health hazards from pollution and poor sanitation. A large percentage of those living in informal housing are migrant workers, especially those from Cambodia and Myanmar. They are unable to obtain even the most basic formal rental accommodation, let alone home ownership, because to their fragile legal position and poor salaries. The city's spatial inequity and planning incoherence are highlighted by the contrast between speculative luxury developments and unofficial encampments.

The conflict between urban growth and ecological preservation is particularly noticeable in Songkhla around Songkhla Lake and the Tinsulanonda Bridge corridor. In order to create space for residential communities, industrial estates, and road infrastructure, wetlands and mangrove forests in Phatong Subdistrict and the southern lake basin have been gradually destroyed. Songkhla's development is more gradual and dispersed than Pattaya's overt high-rise growth or Bangkok's sprawling infrastructure, yet it is just as detrimental to ecological systems.

The growth of informal settlements into buffer or previously preserved areas is a major cause of ecological displacement in Songkhla. Both rural migration and generational land use without official titling are frequently the causes of these settlements. These communities frequently existed before the zoning laws that currently designate their properties as conservation areas, which results in ongoing legal ambiguity and paralysis of enforcement. With an average yearly loss rate of 1.8%, satellite data shows that almost 900 hectares of the mangrove and lowland wetland region surrounding Songkhla Lake were transformed (Dong et al., 2021; Srichaichana et al., 2019) into urban usage between 2005 and 2022.

Ecological displacement is a socio-political phenomenon in addition to a biophysical one. Communities like fishermen and small-scale aquaculture operators that rely on natural ecosystems for their livelihoods experience both spatial marginalization and environmental degradation. This is especially evident in Songkhla, where ancient stilt house villages and lagoon-side markets that served as both cultural and economic hubs are now eroding. Traditional access points to the lake have been obstructed by new real estate projects, and public areas have been replaced by privately owned facilities like marinas or resorts.

The absence of open and inclusive planning procedures exacerbates land-use conflict in all three cities. When necessary, environmental impact studies are frequently carried out following a land acquisition. Weak institutional frameworks, centralized authority, and private-sector influence in land governance sometimes hinder local resistance, especially from informal communities, fisher groups, or environmental activists. Furthermore, communities are left vulnerable to cycles of eviction, poverty, and risk exposure since displacement is rarely supported by effective compensation or relocation plans.

Urban development priorities must be reevaluated in order to address these issues. Informal settlements must be recognized by urban policy not just as issues but also as a component of the urban fabric that merits integration, improvement, and investment. In order to protect delicate ecosystems from irreparable harm, ecological zoning and land-use regulations must be enforced more strictly. A way forward is provided by the incorporation of environmental justice frameworks, which connect ecological preservation with just urban rights. Coordination of institutional transformation, inclusive planning procedures, and consistent investment in ecological infrastructure and safe, reasonably priced housing choices would be necessary to achieve this.

Finally, ecologies being uprooted and informal urbanism becoming more prevalent are not separate problems; rather, they are interconnected signs of a development paradigm that places a low priority on environmental resilience and social justice. Bangkok, Pattaya, and Songkhla's experiences show that resolving land-use conflicts in the context of climate change calls for more than just technical fixes; it also calls for institutional accountability, political will, and an inclusive and sustainable urban future vision.

4.5 Urban Governance and Planning Shortfalls

The governance and urban planning frameworks that influence the management of coastal climate risks are criticized in this section. It looks at fragmented authorities, inconsistent policies, and the lack of local voices in adaptation planning. In Thailand's fast-growing cities, institutional fragmentation, inconsistent policies, and discriminatory planning practices continue to severely limit the ability of municipal government systems to address coastal climate hazards. Although technical evaluations and frameworks for climate adaptation are becoming more widely accessible, there is still uneven implementation of these instruments in Bangkok, Pattaya, and Songkhla in terms of efficient and just governance. From clientelist politics and administrative fragmentation to centralized decision-making and regulatory inertia, these three urban contexts exhibit unique yet connected patterns of governance failure that impede the development and application of successful climate resilience strategies.

Although Bangkok's governance systems are officially centralized under the Bangkok Metropolitan Administration (BMA), district-by-district enforcement varies. Despite the city's access to national adaptation plans, financial resources, and a wealth of technical competence, real implementation is hampered by bureaucratic overload and institutional silos. There are differences between city-wide plans and local execution because of the vertical integration of planning procedures, particularly in land use and water management, which does not adequately coordinate with district offices that are dispersed horizontally. Bangkok's Master Plan on Drainage and Flood Protection, for example, describes advanced modelling of flood-prone areas and infrastructure investment zones; however, district-level adherence to these directives is inconsistent and frequently impacted by capacity constraints, private interests, or political negotiations.

Furthermore, rather than being proactive, Bangkok's municipal governance is becoming more reactive. Long-term structural interventions that address the underlying vulnerabilities of land use and drainage integration are preferred over emergency solutions to flood disasters, such as the use of mobile pumps or canal dredging. The extent and impact of the use of participatory techniques, including local risk mapping or community consultations, are still restricted. Despite being disproportionately impacted by climate-related hazards, low-income and informal populations are frequently left out of conventional planning channels. Their exclusion results in urban adaptation that perpetuates inequalities and loses out on important chances to integrate grassroots knowledge.

The governance shortcomings of Pattaya reveal a new framework, one in which short-term political and economic demands supplant local planning power. Although Pattaya has a semi-autonomous administrative structure as a special municipal area under Chonburi Province, it lacks a coherent development policy and strong institutional leadership. The tourism and real estate industries have a significant impact on urban planning, which leads to quick permits for high-rise complexes, beachside infrastructure, and hillside constructions. These procedures frequently take place without a thorough environmental assessment or hazard zoning compliance. Land-use laws are routinely disregarded, changed, or circumvented, especially when there is a lot of political upheaval at the local level.

In Pattaya, ad hoc planning impedes the development of resilience by sustaining land-use incoherence. For instance, in later years, stormwater retention sites that were authorized in the 2010 urban plan were rezoned for commercial use, frequently without informing the public. Similarly, when they are carried out, environmental impact studies for significant constructions are rarely incorporated into infrastructure plans or citywide adaptation strategies. Long-term climate governance is further hampered by the absence of institutional continuity. A fragmented development landscape with little accountability results from new administrations frequently reversing or stalling the projects started by their predecessors. There is very little civic engagement and almost no system in place for community participation in environmental monitoring or land-use choices.

Poor inter-agency collaboration and administrative inertia have shaped Songkhla's governance environment. The city has a structural gap between environmental management authorities and urban planning institutions, despite being smaller and less politically unstable than Bangkok or Pattaya. While distinct provincial and national organizations regulate water resources, coastal ecosystems, and catastrophe risk, the local municipal government is in charge of zoning, public works, and infrastructure development. Particularly when it comes to carrying out integrated climate adaptation programs, this jurisdictional split encourages a lack of accountability and cooperation.

Due to overlapping responsibilities and a lack of enforcement capacity, the growth of residential and industrial zones into environmentally vulnerable areas surrounding Songkhla Lake has gone unchecked. In the absence of legally binding frameworks or municipal collaboration, environmental protection authorities may highlight problems associated with water pollution or mangrove damage, but these issues are frequently ignored. Furthermore, Songkhla's growth plans are still mainly antiquated and static, lacking both forward-looking vulnerability assessments and real-time climatic data. Budgetary restrictions and administrative hold-ups have thwarted efforts to implement ecosystem-based adaptation strategies.

Governance deficiencies are made worse in all three cities by the lack of effective public participation tools. Stakeholder consultation, decentralization, and environmental hearings are all theoretically covered by Thailand's planning laws. In reality, these procedures are frequently hurried, tokenistic, or circumvented. Land-use policies and infrastructure objectives are rarely shaped by university researchers, urban poor populations, or environmental advocates. As a result, choices impacting thousands of vulnerable individuals are made without their knowledge or input, creating a democratic deficit in urban climate governance.

Furthermore, Thailand's urban development political economy exacerbates governance deficiencies. Municipal planning agendas are frequently dominated by real estate interests, since zoning and investment decisions are influenced by elite alliances, land speculation, and campaign funding. These dynamic distances climate governance from public accountability and erodes regulatory integrity. When infrastructure is invested in, economic expansion is typically prioritized over social protection or environmental resilience. Green infrastructure, for instance, is often underfunded and politically deprioritized relative to more visible and commercially valuable assets like roads, convention centers, or tourist amenities.

Urban institutional frameworks must be reconfigured to address these planning and governance deficiencies. Bangkok needs more robust horizontal coordination systems that provide local officials with tools and directives in line with the city's climate goals. Pattaya requires stronger environmental control, zoning integrity, and regulatory protection against political upheaval. In order to facilitate integrated decision-making that safeguards its vital ecosystems, Songkhla has to close the gap between planning and environmental governance.

Building equitable and successful urban resilience in all situations requires improving participatory governance, which can be achieved through community climate councils, co-production of risk data, and civic accountability systems. Urban climate governance is more than just crisis management or technical planning. Power, inclusivity, and the ability to envision and institutionalize many urban futures all influence this political process. If Thailand's coastal cities want to handle climate risks in a fair and efficient manner, they must face these realities.

4.6 Toward Nature-Positive Urban Transformation

This final section identifies emerging practices and future opportunities for embedding ecosystem-based adaptation and climate-sensitive design into urban planning across the three cities. It promotes participatory governance, integrated planning, and urban ecologies restoration. Reorienting urban development techniques from reactive engineering to anticipatory, ecological, and inclusive urbanism is essential as Thailand's coastal cities deal with the growing effects of climate change. This chapter has demonstrated how land-use conflicts, fragmented governance, and uneven infrastructure, in addition to geographic vulnerability, exacerbate climate risks in Bangkok, Pattaya, and Songkhla. Nevertheless, there are also new signs of change inside this intricate environment of danger. These chances provide a route to nature-positive urbanization, where ecological restoration, climate resilience, and community well-being are all pursued as interrelated objectives.

In Bangkok, the extent of infrastructure deterioration and urbanization has coincided with an increasing understanding of the potential of blue-green networks for urban adaptability. Initiatives like the greening of Ratchadamnoen Avenue and Chulalongkorn University's Centenary Park provide modest but calculated moves toward incorporating water-sensitive design into the spatial logic of the city. Once written off as development reserves, Bangkok's peri-urban areas have enormous potential as ecological buffers. Along the eastern floodplain, retaining rice paddies, canal systems, and green corridors might have two purposes: they could act as local food systems and flood retention areas. Instead of being surrendered to speculative real estate, these peri-urban agricultural landscapes can be protected and improved as climatic assets through the implementation of smart zoning reforms and co-financing methods.

In contrast, Pattaya has to more openly address the conflict between ecological deterioration and economic viability. Even while its tourism industry is thriving, it cannot last if it keeps threatening the coastal systems that it depends on. The city's economic basis might be diversified while its environmental capacity is restored by moving toward low-impact tourism zoning, which would prioritize eco-resorts, attractions related to conservation, and the conversion of degraded areas into community-managed green spaces. Replanting mangroves along coastal fringes provides a useful and symbolic anchor for this transformation, especially in the underdeveloped areas of Jomtien and Naklua. Here, pilot projects have the potential to spark more extensive community involvement and offer green infrastructure with quantifiable co-benefits for carbon sequestration, biodiversity, and storm protection.

The potential in Songkhla is found in integrated hybrid approaches that combine ecosystem-based adaptation with hard infrastructure. Songkhla may take the lead in creating nature-based solutions that are suited to mid-sized, ecologically rich communities rather than depending only on dilapidated drainage enhancements or deteriorating seawalls. When combined with eco-corridors that connect urban areas to the lake's edge, wetland regeneration surrounding Songkhla Lake can lower runoff, enhance water quality, and reestablish traditional ties to aquatic life. Rehabilitated mangrove belts, fish sanctuaries, and community-managed buffer zones can serve as the cornerstone of a local resilience economy that is based on stewardship rather than extraction.

In all three cities, institutional bravery and inclusive governance—rather than technological innovation—are the most revolutionary components. Ecological thinking can be mainstreamed into regular urban decision-making by incorporating nature-positive planning into statutory tools, participatory budgeting procedures, and educational programs. Cross-sectoral cooperation, persistent political determination, and an understanding that resilient futures are based on repair and reimagining rather than relocation and denial are necessary for this.

Nature-positive urbanism is not a utopian ideal, it is a necessity for cities at the edge. At this point, Bangkok, Pattaya, and Songkhla must decide whether to continue on their current course of ecological degradation and piecemeal adaptation or to daringly reimagine their futures using inclusive, regenerative, and climate-sensitive methods. The seeds of this change are already in place. Developing them into a robust urban environment for future generations is the challenge.

References

Allam, Z., & Cheshmehzangi, A. (2024). *Sustainable futures and green new deals* (pp. 1–137). Palgrave Macmillan.

Anukularmphai, A., Sompongchaiyakul, P., Rivas, A., & Sangganjanavanich, P. (2005). *A case study for interlinked coastal and river basin management for the Songkhla Lake Basin, Thailand.*

Cheshmehzangi, A. (2023). Regenerating local industries and urban experiences that matter. In *Mapping urban regeneration: City life experiences in Yunnan, China* (pp. 23–75). Springer Nature Singapore.

Dong, K., Qiao, S., Wu, B., Shi, X., Chen, Y., Shan, X., Liu, S., Kornkanitnan, N., & Khokiattiwong, S. (2021). Sedimentary history of trace metals over the past half-century in Songkhla Lake, western coast of the Gulf of Thailand: Anthropogenic impacts and contamination assessment. *Frontiers in Earth Science, 9*, Article 767899.

Huang, H. O. (2019). *Research of the tourism industry development strategy in Bangkok of Thailand* (Unpublished master's thesis). Siam University, Thailand.

Jongkroy, P., & Thongbai, C. (2019). The potentials and limitations, and specialization of regional cities in Thailand. *Kasetsart Journal of Social Sciences, 40*(2), 440–447.

Menakanit, A., Davivongs, V., Naka, P., & Pichakum, N. (2022). Bangkok's urban sprawl: Land fragmentation and changes of peri-urban vegetable production areas in Thawi Watthana district. *Journal of Urban and Regional Analysis, 14*(1), 59–78.

Palakhamarn, T., & Kamolvej, T. (2024). Revitalizing urban resilience in Thailand: Exploring conceptual frameworks and terminology. *Journal of Architectural/planning Research and Studies (JARS), 21*(2), 263–282.

Protong, S. (2016). *Climate change and landslide risk assessment in Thailand* (Doctoral dissertation, University of Southampton).

Sahacharoenwat, P. (2024). *Understanding the complex dynamics of regenerating urban vacancy: A case study of Songkhla, Thailand* (Doctoral dissertation, Massachusetts Institute of Technology).

Srichaichana, J., Trisurat, Y., & Ongsomwang, S. (2019). Land use and land cover scenarios for optimum water yield and sediment retention ecosystem services in Klong U-Tapao Watershed, Songkhla, Thailand. *Sustainability, 11*(10), 2895.

Tabucanon, M. S. (2004). *Governance in natural resources and environmental management in Songkhla Lake Basin* (p. 62). Rule of Law for Good Environmental Governance.

Tontisirin, N., & Anantsuksomsri, S. (2021). Economic development policies and land use changes in Thailand: From the eastern seaboard to the eastern economic corridor. *Sustainability, 13*(11), 6153.

Wang, Y., Hu, Y., Niu, X., Yan, H., & Zhen, L. (2022). Land use/cover change and its driving mechanism in Thailand from 2000 to 2020. *Land, 11*(12), 2253.

Yokota, S., Aihara, Y., & Thaitakoo, D. (2020). Land use transition and possibility of climate change adaptation by rice farmers around the industrial estate in Bangkok's urban fringe. *Urban and Regional Planning Review, 7*, 22–42.

Chapter 5
Locally Anchored Nature-Based Solutions in Coastal Thailand: Ecosystem-Based Adaptation in Si Chang Island and Phetchaburi

5.1 Introduction: Coastal Ecosystems Under Pressure

Setting the scene, this first segment looks at how climate change, fast development, and dwindling biodiversity are making Thailand's coastal zones more ecologically and socioeconomically vulnerable. It presents the case for nature-based solutions (NbS) and ecosystem-based adaptation (EbA), highlighting their applicability to communities such as those in Phetchaburi and Si Chang Island.

With a variety of marine life, lush estuaries, and people reliant on fishing, aquaculture, salt farming, and tourism, Thailand's coastal regions are both environmentally and economically significant (Bennett et al., 2014; Ruyabhorn & Phantumvanit, 1988). However, climate change, increased urbanization, unsustainable land use, and environmental degradation are now putting increasing strain on these same coastlines. Developmental pressures like port expansions, coastal erosion, plastic waste, and overexploitation of marine resources are exacerbating the physical effects of sea level rise, stronger storms, and salt intrusion. These stresses are actual realities rather than hypothetical possibilities in locations like Si Chang Island and the Phetchaburi region, where they show up as diminishing fish populations, coral bleaching, the loss of mangrove buffers, and more precarious local livelihoods.

Thailand has over 3200 kms of coastline stretching across the Gulf of Thailand and the Andaman Sea, encompassing a diversity of ecosystems including mangroves, seagrass beds, coral reefs, salt marshes, estuaries, and low-lying deltas. These ecosystems provide invaluable services like storm surge protection, fishery support, carbon sequestration, water quality regulation, and food and livelihood security. They are, therefore, also among the most susceptible to risks associated with climate change. Over the last 20 years, coastal erosion rates have significantly increased, particularly in the Central and Upper Southern regions. Once common, mangrove forests have been weakened by infrastructural development, land reclamation, and shrimp farming. Warming sea temperatures and sedimentation from land-based activities are

A. Cheshmehzangi, *Thailand's Tropical Thresholds*, Urban Sustainability,
https://doi.org/10.1007/978-981-95-4975-7_5

causing bleaching and die-off in seagrass meadows and coral reefs, which are vital habitats for marine species.

This ecological pressure's socioeconomic component is also worrying. Informal tourism operators, smallholder salt producers, and rural fishing communities frequently lack the institutional and financial resources necessary to adjust to shifting environmental conditions. Furthermore, a large number of these communities function at the outskirts of official planning frameworks, having little access to measures for protecting the coast or involvement in environmental governance. Small-scale fishermen on Si Chang Island, for instance, are finding it more and more difficult to balance commercial maritime operations with marine conservation regulations. Communities that depend on salt farming in Phetchaburi are dealing with the strains of land conversion as well as unpredictable weather patterns that interfere with harvesting seasons.

Against this backdrop, the necessity of adaptive measures that both lessen vulnerability and rebuild the natural underpinnings of coastal resilience is becoming increasingly apparent. Here, nature-based solutions (NbS) and ecosystem-based adaptation (EbA) have become important concepts. In order to manage climate and development risks, both strategies place a focus on working with natural systems rather than against them. EbA is a subset of NbS that focuses on climate resilience through ecosystem services, whereas NbS is a more general word that includes biodiversity conservation, disaster risk reduction, and climate adaptation.

The promise of EbA and NbS is found in their capacity to produce co-benefits, which include lowering physical exposure to risks and improving local livelihoods, biodiversity, and cultural identity (Fan & Cheshmehzangi, 2022). They offer an alternative to concrete infrastructure like dikes and sea barriers, which may give protection right away but over time frequently weaken ecological systems and lead to maladaptive dependencies. Replanting mangroves in storm-prone locations, for example, provides long-term resilience that changes with the ecosystem by stabilizing silt, restoring fish habitats, and buffering wave energy.

These strategies have been gaining popularity in Thailand, especially through community-led campaigns and donor-funded initiatives/projects. The integration of EbA into Integrated Coastal Zone Management (ICZM) frameworks (Nagabhatla et al., 2018) has been investigated in projects funded by the UNDP, local NGOs, and the Green Climate Fund (GCF) (Vikas & Hari, 2023). However, many communities have long-standing customs that support the EbA ideals, like rotating harvesting, preserving holy groves, and restricting fishing during certain seasons. These systems of traditional ecological knowledge (TEK) (Kim et al., 2017; Martin et al., 2010) can be combined with modern sustainability research to create a potent platform for place-based adaptation and innovation.

However, the success of NbS and EbA is not guaranteed. Lack of long-term finance, tokenistic planning, poor monitoring, and a propensity to treat them as stand-alone initiatives rather than essential components of development and spatial planning frequently hinder their implementation (Li et al., 2021). Moreover, without deep community engagement, there is a risk of technocratic imposition, where external actors define problems and solutions without fully grasping local realities. Therefore, maintaining and expanding nature-based adaptation requires locally grounded strategies that are founded on lived experience and co-developed with communities.

This chapter places itself at the nexus of these opportunities and difficulties. It aims to provide light on how NbS and EbA are being used, modified, and recreated in two very different coastal environments by looking at case studies from Si Chang Island and the Phetchaburi area. Through zoning, coral reef restoration, and participatory marine spatial planning, the little island of Si Chang, located off the eastern seaboard, provides insights into the recovery of marine ecosystems. With its vast salt flats and lowland estuaries, Phetchaburi serves as an example of how biodiversity-friendly farming methods and mangrove buffer regrowth may coexist in dynamic deltaic environments.

These case studies collectively advance our knowledge of what it means to embed nature-based climate adaptation in the uniqueness of location, where social behaviours, ecological functions, and governance capacities come together. They also highlight the conflicts between localization and expansion, between community implementation and policy ambition, and between economic motivations and ecological reasoning. The urgency is obvious as Thailand continues to struggle with the twin demands of economic growth and environmental change: ecosystems, equity, and empathy must be used in addition to concrete and capital to build climate resilience. The next sections examine the ways in which two representative coastal regions are cultivating, contesting, and co-creating such resilience.

5.2 Case Study I—Si Chang Island: Seagrass Recovery and Marine Zoning

An excellent example of ecosystem-based adaptation in a tiny island context where economic, ecological, and geopolitical forces combine is Si Chang Island (Fig. 5.1), which is situated in Chonburi Province off the eastern coast of Thailand (Chirapart et al., 2003). The island, which is situated along marine trade lanes and is expanding under the influence of both industrial and tourism, is an important node in the Gulf of Thailand despite its relatively modest landmass. In spite of this complexity, local actors have started creative approaches to community-led conservation and the restoration of marine ecosystems (Umprasoet et al., 2023), especially in the areas of seagrass recovery, coral regeneration, and adaptive marine zoning.

Historically, a variety of marine life, such as dugongs, sea turtles, and many fish species, have been sustained by the coastal and subtidal habitats of Si Chang. However, increased boat traffic, overfishing, sedimentation from shoreline building, and nutrient loading from tourism-related activities have all contributed to the island's notable ecological decline during the last 20 years. Seagrass meadows, which act as nurseries for young fish and offer vital habitat connections in the Gulf, have been among the most obviously impacted environments (Kritsanapuntu et al., 2024). In

Fig. 5.1 Si Chang Island and its proximity to mainland regions of Chonburi Province, also showing the extended economic and tourism activities (*Source* Map generated by the author using Google Earth Pro, from July 2025 maps of the region—openly available by 2025 Maxar Technologies and 2025 Airbus)

addition to endangering biodiversity, the loss of seagrass beds also damages the ecological services—artisanal fishing and ecotourism—that sustain local livelihoods.

As a result, community organizations have started seagrass restoration initiatives utilizing species that are peculiar to a certain location, like Halophila ovalis and Enhalus acoroides, with assistance from marine biologists and local government entities (Bujang et al., 2016). In an effort to instil a conservation attitude in future generations, these programs are sometimes paired with environmental education initiatives in nearby schools. The synergistic way that traditional ecological knowledge is being re-integrated into modern restoration efforts, especially among senior fishermen, is noteworthy. Historically prolific seagrass zones, the best times of year to plant, and regional markers of ecosystem health like species return and water clarity have all been identified with the assistance of fishermen.

These ecological restoration initiatives have aligned with new maritime zoning regulations at the governance level, which aim to lessen human impact in environmentally vulnerable areas. These frameworks use ecological sensitivity and community consultation to define areas for various activities, such as shipping, fishing, tourism, and conservation. Community input has influenced changes over time in this experimental and iterative zoning process. In addition to halting more deterioration, this spatial difference has made it possible to allocate funds specifically for ecosystem monitoring (Shoo et al., 2021) and recovery.

The island's comparatively modest size has helped to promote these coordinated strategies. Regular communication, adaptive learning, and social accountability are made possible by the close proximity of stakeholders, including local officials, fishermen, tourism operators, and school administrators. Additionally, university collaborations have offered technical assistance for co-designing policies and monitoring, especially with Gulf-based marine research institutes. But there are still difficulties. Implementation has occasionally been hampered by jurisdictional overlaps between marine departments and municipal authorities, and community-led projects' ability to continue is limited by a lack of sustained funding.

The natural benefits—like the partial regeneration of seagrass beds and the rise in marine biodiversity—as well as the governance improvements brought about by place-based experimentation are what make Si Chang an especially instructive case. The design and effects of the coral-friendly marine zoning and community participation mechanisms that support these accomplishments are covered in more detail in the next subsection.

5.2.1 *Coral-Friendly Marine Zoning and Community Engagement*

Si Chang Island's marine zoning is an example of a grounded, flexible experiment in spatial government that puts community involvement and ecological integrity ahead of top-down command-and-control methods (Umprasoet et al., 2023). The zoning process, which is founded on the idea of ecosystem-based management, started because of locally perceived needs such as diminishing fish stocks, damaged coral reefs, and disputes between boat operators, divers, and fishermen. Instead of addressing these problems separately, local government and community leaders looked for a spatial plan that would reorganize maritime space use to conform to social norms and environmental standards.

Fishermen, tour guides, conservation NGOs, educators, and local regulators participated in multiple rounds of participatory mapping workshops that resulted in the creation of the coral-friendly maritime zoning plan, which has been piloted for the past five years. Stakeholders identified important ecological zones, historic fishing grounds, popular tourist destinations, and recognized coral beds using basic resources including printed nautical maps, coloured overlays, and post-it notes. A dynamic spatial plan was created by digitizing these maps and overlaying them with satellite data and ecological surveys. Four main areas were identified: (1) areas for the protection of coral reefs, (2) areas for artisanal fishing, (3) areas for ecotourism, and (4) areas for general usage.

The prohibition of motorized boat anchoring in shallow reef zones, which had previously led to physical coral damage and sediment disruption, was one of the first and most significant zoning effects. In partnership with regional university students and local dive operators, floating buoys were installed to provide a non-intrusive mooring option that improved diver safety and protected coral at the same time. To allow reef species to repopulate and stabilize ecosystem functioning, fishing in regions adjacent to coral was temporarily prohibited. Peer monitoring and community pride served as the main enforcement mechanisms for these closures, which were supported by unofficial social contracts rather than legal sanctions.

The zoning strategy's success has been largely attributed to the involvement process. Marine conservation has been explained, promoted, and celebrated through town hall meetings, village radio broadcasts, youth-led campaigns, and even island festivals. As "reef ambassadors", local schoolchildren lead snorkelling excursions and engage in citizen science projects that strengthen the transfer of knowledge between generations. Crucially, these programs have presented conservation as a means of maintaining livelihoods rather than as a trade-off against them. In some locations, fishermen are now reporting larger catches, and ecotourism businesses are starting to promote the island's "restored reef experience" as a differentiator.

There are still difficulties. Zone boundaries can occasionally lead to conflicts, especially between conservationists and operators of informal tourism. Furthermore, there is continuous conflict with marine agencies at the national level, whose policy directives don't always coincide with regional spatial interests. But these conflicts have also given rise to innovative platforms for negotiations. For example, monthly gatherings called the "Marine Monday Dialogues", which are co-hosted by university marine scientists and local leaders, have developed into a forum for converting scientific information into decisions that affect the community.

The Si Chang experience demonstrates that coral-friendly zoning is a process of community reorganization, ecological stewardship, and adaptive governance rather than merely a geographical exercise. It demonstrates how even small island communities may develop marine governance policies that are both successful and firmly established when local knowledge, participatory planning, and ecosystem research are combined.

5.3 Case Study II—Phetchaburi Region: Mangrove Restoration and Estuarine Resilience

Rich estuarine habitats, complex canal networks, and vast coastal lowlands define the Phetchaburi region, which is situated on the upper Gulf of Thailand (Fig. 5.2). This region has long sustained livelihoods that are closely linked to the natural world, including artisanal fishing, aquaculture, and small-scale salt farming. But the area is now more in the forefront of ecological fragility. The estuaries' delicate salinity gradients have been upset by upstream water diversions, saline incursion, and sea level rise (Pumijumnong, 2013). Concurrently, the region's ecological equilibrium has been further upset by the conversion of mangrove forests for tourist facilities and shrimp farms.

Fig. 5.2 The other side of the upper Gulf of Thailand in Phetchaburi region and its vast coastal lowlands (*Source* Map generated by the author using Google Earth Pro, from July 2025 maps of the region—openly available by 2025 Maxar Technologies and 2025 Airbus)

In the past, Phetchaburi's mangroves served as an uninterrupted green wall separating the sea from populated areas. These ecosystems supported soil fertility and controlled microclimates in the salt flats in addition to protecting the coast and offering vital homes for marine life. Much of the coastline is now vulnerable to erosion and storm surges as a result of the extensive clearing of these woods in the late twentieth century, mostly for the expansion of aquaculture. Additionally, the loss of vital fish and crustacean nursery grounds, increased sedimentation in canals, and increased susceptibility to tidal floods have resulted from the disappearance of mangrove buffers.

However, mangrove restoration has become increasingly important as a key component of ecosystem-based adaptation in the area in recent years. A combination of scientific enlightenment, policy reorientation, and—above all—local community mobilization has sparked this change (Bidorn et al., 2020). Numerous community organizations, frequently made up of retired fishermen and salt farmers, have spearheaded efforts to restore damaged mangrove areas with natural species like Avicennia marina and Rhizophora apiculata. These initiatives are frequently carried out in association with environmental NGOs, sub-district administrative organizations, and nearby colleges.

The removal of invasive species and detritus is usually the first step in restoration efforts. Next, shoreline conditions are stabilized using natural materials like coconut husk barriers or bamboo fences. Mangrove seedlings are planted in staggered rows to replicate natural patterns after sediment dynamics have been restored. Crucially, these reforestation initiatives are intended to be socio-ecological interventions as well as environmental programs. To ensure cultural resonance and wider participation, many planting efforts are scheduled in conjunction with local festivals or religious rituals.

Concurrently, there is a revived interest in regulating land and water uses throughout the coastal-inland gradient because of growing understanding of the significance of estuarine integrity. Farmers are experimenting with more salt-tolerant crop types or switching to integrated aquaculture systems because of saline intrusion rendering certain inland locations unsuitable for conventional rice growing. As a result, the landscape is undergoing a subtle but important reconfiguration, where adaptation involves striking a new balance between environmental change and agricultural viability rather than merely going back to an earlier biological state.

Through funding, training, and official recognition initiatives, government organizations like the Department of Marine and Coastal Resources (DMCR) have been supporting these activities more and more. Under Thailand's National Adaptation Plan, Phetchaburi has been incorporated into regional adaptation planning, namely under the framework for Coastal Resources and Ecosystem Resilience. Additionally, grants for water management, biodiversity conservation, and climate-resilient livelihoods in estuarine areas have been made available by international agencies like the GCF.

Yet, challenges persist. Conflicts over commercial development plans or land tenure are preventing certain places designated for mangrove restoration from moving forward. Despite their economic significance, shrimp farms still invade restoration areas. Furthermore, it is challenging to evaluate the actual effects of restoration initiatives on ecosystem health and community well-being due to the absence of regular long-term monitoring. Institutional weariness is another issue that many communities deal with, as short project cycles and conflicting requirements lessen the motivation for ongoing involvement.

Nonetheless, the story of adaptation in Phetchaburi continues to revolve around the tenacity of community actors and their inventiveness in real-world situations. When planning is inclusive and flexible, local innovations—like utilizing estuarine waterways for multi-species aquaculture or combining mangrove planting with artisanal salt crystallization platforms—show that ecosystem-based solutions may coexist alongside commercial output. The following subsection delves deeper into these hybrid land-use patterns, specifically focusing on how community-driven design is integrating agro-ecological revitalization and biodiversity protection (Tansuchat & Plaiphum, 2023) into salt-farming landscapes.

5.3.1 Adaptive Land Use and Agro-ecological Revitalization

Restoring destroyed ecosystems is only one aspect of adaptation in Phetchaburi's dynamic coastal landscapes; another is rethinking land use in ways that revitalize both ecological processes and rural livelihoods. Particularly in the context of traditional salt farming, which has long influenced the region's physical and cultural landscape, this subsection focuses on how people are innovating at the nexus of biodiversity conservation and agro-ecological resilience (Cheshmehzangi, 2022).

In Phetchaburi, salt farming is a seasonal endeavour that is mostly dependent on the interaction of the sun, wind, and the availability of brine from neighbouring estuary systems. Many salt producers have seen decreasing yields and increased economic uncertainty because of changing climate patterns and unpredictable water regimes. Some communities have responded by adopting integrated land-use patterns that incorporate biodiversity corridors, aquaculture ponds, ecological buffers, and salt crystallization platforms. With these hybrid systems, monocultural land use is giving way to multipurpose landscapes that improve resilience.

Reserving the outer rows of salt plots for the planting of mangroves or nipa palms is a commonly used technique that reduces saline creep into nearby farmlands while generating windbreaks and sediment traps. Farmers have also set up raised garden beds in flood-prone areas to cultivate salt-tolerant herbs with culinary and therapeutic uses, like Pandanus and Centella asiatica. Small-scale circular water management systems are made possible by the frequent use of collected rainfall or filtered estuary discharge to irrigate these beds.

One especially effective example is the Ban Laem sub-district, where a group of salt producers established a "Salt-Forest Mosaic" project in partnership with a nearby environmental non-governmental organization (Thiammueang et al., 2014). As part of the project, areas of underutilized salt flats were delineated for biodiversity regeneration, native halophytes were planted, and aquatic and migratory bird micro-habitats were established. The resulting landscape serves as a barrier against coastal flooding, preserves species habitat, and produces seasonal revenue from salt, making it a system that is both protective and productive. Additionally, ecotourism activities have been incorporated, with guests being carried on elevated wooden pathways around the mosaic and learning about the connections among ecosystem services, climate adaptation, and traditional knowledge.

These methods are rooted in deeply place-based knowledge systems. While young organizations have taken the lead in biodiversity mapping and social media outreach, elders frequently serve as guardians of seasonal calendars and salt-making customs. This combination of digital communication and intergenerational learning has increased the movement's prominence and drawn outside funding for expansion. Additionally, collaborations with regional institutions have improved local decision-making by introducing ecological assessment tools including participatory GIS mapping, drone-based land monitoring, and salinity sensors.

Local development plans have been modified by certain sub-district administrative organizations to specifically incorporate ecosystem-based land use zones at the governance level (Kitjarak et al., 2024). Although it is still in its infancy, this change indicates that agro-ecological design is now acknowledged as a valid adaptation strategy. As long as institutional frameworks are made simpler and local perspectives are prioritized during the planning process, it also creates room for climate funds to be directed directly to community-led projects.

However, a number of obstacles must be overcome to maintain these efforts. Obstacles still include short finance cycles, a lack of technical expertise, and unclear land rights. Due to perceived efficiency risks, commercial salt operations in some regions—particularly those connected to export markets—resist integrating conservation aspects. Furthermore, there is still worry about youth outmigration, which jeopardizes community cohesion and the continuation of customs.

Notwithstanding these obstacles, Phetchaburi's adaptive land-use changes show how low-tech, socially entrenched, and highly contextualized solutions can promote resilience. These programs represent climate adaptation as a lived, changing practice that is influenced by ecological understanding, economic necessity, and shared care for land and water, rather than viewing it as an external imposition.

5.4 Scaling Up Nature-Based Adaptation: Opportunities and Challenges

The Phetchaburi region and Si Chang Island serve as examples of how ecosystem-based adaptation (EbA) and nature-based solutions (NbS) can be successfully rooted in local ecological knowledge, community values, and context (Walz et al., 2021). Prospective adaptation approaches that are socially inclusive, ecologically regenerative, and functionally integrated into local lives are reflected in these bottom-up projects. However, it is still very difficult to translate these place-based initiatives into scalable frameworks that might influence more general development and climate agendas throughout Thailand's coastal regions. The prospects and difficulties of scaling up NbS are critically assessed in this section, with an emphasis on the institutional setups, financing sources, enabling circumstances, and policy processes needed to mainstream these strategies.

The understanding that effective NbS therapies are very place-dependent lies at the heart of the scaling debate. They rely on trust developed over time, informal governing networks, and deep understandings of local ecologies. Therefore, contextual adaptation, co-design, and socio-ecological fit must be given priority in any endeavour to reproduce or scale such models rather than just repeating activities. Si Chang and Phetchaburi both provide valuable insights in this area. The participatory marine zoning technique was successful in Si Chang not because of a strict template but rather because it enabled user groups to negotiate iteratively, combining community reasoning with scientific input. The persistence of hybrid salt-mangrove systems in Phetchaburi was a result of multilayer adaptive techniques, flexible resource usage, and intergenerational land knowledge.

However, institutional frameworks that respect and codify local experimentation are necessary for these techniques to transcend isolated pilots and affect national or regional policy. Funding sources that are adaptable, long-term, and available to grassroots actors are the first step in this process. In Southeast Asia, including Thailand, ecosystem-based adaptation is already being supported by international climate finance tools like the Global Environment Facility (GEF), Adaptation Fund, and Green Climate Fund (GCF) (Cui et al., 2020). These grants frequently direct resources through national organizations that subsequently oversee provincial or local implementation. Although this method provides macro-level alignment with sustainable development goals (SDGs) and national adaptation plans (NAPs), it usually ignores the fine-grained, slow-building processes that support community-led adaptation.

The adoption of "enhanced direct access" models, in which monies are made accessible directly to local governments or recognized community organizations, is one solution that is gaining popularity. This would entail giving provincial-level environmental committees or sub-district administrative organizations (SAOs) the authority to actively propose, carry out, and oversee EbA projects in Thailand. Compared to centralized ministries, these actors frequently possess a deeper comprehension of local ecological processes and stakeholder landscapes. Direct access mechanisms can democratize adaptation finance while maintaining the local agency that is essential to the success of NbS when combined with venues for knowledge-sharing and capacity-building.

However, without supportive policy frameworks, financial support is insufficient on its own. Thailand's coastal planning is still dispersed throughout several departments and organizations, from provincial land use boards and tourism authorities to the Royal Irrigation Department and the Department of Marine and Coastal Resources (DMCR). Each has its own goals, measurements, and timeframes for planning. In addition to creating competing land and water use goals, this fragmentation marginalizes NbS's integrated, cross-sectoral rationale.

In Phetchaburi, for instance, community initiatives to reforest mangroves can conflict with plans for the expansion of aquaculture at the provincial level or extensive zoning schemes for tourists. Similar to this, national marine spatial planning schemes may not explicitly acknowledge marine zoning for the protection of coral and seagrass on Si Chang Island, which could lead to difficulties in long-term administration and enforcement. These conflicts highlight how urgently cross-scale cooperation is needed, in which local initiatives are nested inside logical regional frameworks that connect planning tools, administrative borders, and natural systems.

There are various ways to promote policy integration. One is the official incorporation of NbS and EbA into national planning tools including coastal development guidelines, the National Biodiversity Strategies and Action Plans (NBSAP), and the National Adaptation Plan (NAP) (Terton et al., 2024). Although ecosystem-based initiatives are mentioned in Thailand's most current NAP, there is still a lack of operational clarity regarding how these strategies would be financed, carried out, and assessed locally. Integrating EbA into standard planning and monitoring processes requires clearer mandates and metrics that account for co-benefits, including as social fairness, livelihood security, and ecosystem health.

Decentralized planning reform presents yet another significant possibility. Many local governments now lack the power or technical capability to effectively integrate climate adaptation into their financial and geographical planning. When towns are provided with the necessary tools, resources, and institutional liberty to experiment, adaptation can be incorporated into local development plans, as demonstrated by pilot projects in other Southeast Asian nations. Structured support programs could help Thailand's SAOs and municipalities integrate NbS into their local land use plans, infrastructure investments, and disaster risk reduction initiatives, particularly in climate-sensitive coastal districts.

Notwithstanding these routes, a few warnings need to be taken into account. The term "nature-based solutions" can be used to describe actions that are environmentally superficial or solely motivated by political or commercial motivations, a phenomenon known as "greenwashing". For example, creating artificial wetlands for tourism purposes or establishing monocultures in degraded mangrove areas can be presented as NbS, but they may actually reduce biodiversity and community utility. Therefore, a robust framework for monitoring and verification that is based on ecosystem functionality and community benefit is essential.

Second, scaling initiatives need to prevent solution homogenization. There is a propensity to look for standardized templates or toolkits as NbS gains traction in donor and policy circles. This strategy may unintentionally replace regional customs, disregard cultural factors, and enforce success criteria that are determined by outside sources. The distinctiveness of place—the regional rhythms, ecological wisdom, and adaptive behaviours—is what creates resilient results in both Si Chang and Phetchaburi. In this sense, scaling should be viewed as distributed innovation, where fundamental ideas are modified for use in many institutions and environments, rather than as uniform replication.

Lastly, any attempt to broaden EbA must continue to prioritize real community participation. This entails acknowledging local actors as co-creators of knowledge and stewards of implementation in addition to being recipients. It entails creating transparent, deliberative governing systems that include the voices of marginalized groups, including youth, women, Indigenous peoples, and informal workers. Co-design, shared monitoring, and long-term stewardship are other forms of participation that should go beyond the consultation stage.

In conclusion, there are both opportunities and challenges associated with expanding nature-based adaptation in Thailand's coastal zones. Locally anchored approaches could promote ecological regeneration and community resilience, as demonstrated by the success of efforts in Si Chang Island and the Phetchaburi region. However, Thailand must address policy fragmentation, restructure financial routes, decentralize adaptation planning, and develop frameworks that respect local variety if it hopes to achieve this promise on a large scale. Justice, honesty, or specificity must never be sacrificed for scale. A regenerative ethos, which views adaptation as an invitation to co-evolve with the land, the sea, and the people who depend on them rather than as an imposition of paradigms, must instead serve as its compass.

5.5 Conclusion: Toward Embedded, Context-Sensitive Coastal Planning

A crucial reality is highlighted by the experiences of Si Chang Island and the Phetchaburi region: effective climate adaptation in coastal Thailand requires a profound integration of local ecological, social, and cultural factors. When created and used using context-sensitive methods, nature-based solutions (NbS) can transcend environmental interventions and become manifestations of local ingenuity, stewardship, and resilience (Olivadese, 2025). However, they can only reach their full potential if they are viewed as essential elements of the national adaptation and coastal planning agenda rather than as side projects.

These two case studies show how ecosystem-based adaptation (EbA) can function well at the nexus of lived experience and scientific understanding. In addition to biological monitoring and zoning, ongoing communication with fishermen, divers, educators, and local authorities enabled the regeneration of seagrass meadows and coral zones in Si Chang. In addition to a healthier marine environment, the outcome is a sense of collective ownership and co-management. Estuarine landscapes in Phetchaburi are changing in ways that preserve ecological integrity and economic viability through the hybridization of salt farming with mangrove restoration and biodiversity conservation. Knowledge is also co-produced here, arising from the cycles of salt crystallization, memories of flood cycles, and a readiness to modify farming methods to accommodate shifting environmental circumstances.

Thailand's larger coastal planning frameworks can learn a lot from these experiences. The significance of local agency comes first. The most sustainable adaptation initiatives, whether in mangrove restoration or maritime zoning, are those that are the result of community initiative, influenced by traditional knowledge, and moulded by iterative learning. Therefore, it is imperative that coastal planning avoid one-size-fits-all approaches and instead use frameworks that enable local actors to understand, use, and modify NbS in accordance with their unique situation.

Cross-scalar integration is crucial, too. When enabling institutional frameworks that span administrative levels support local initiatives, they thrive. Thailand's maritime spatial frameworks, biodiversity action plans, and national adaptation strategy must make room for locally developed models to scale both vertically—across communities—and horizontally—through incorporation into national and subnational policy tools. Decentralizing planning authority is necessary for this, but so are investments in technical assistance, capacity building, and participatory procedures that provide local governments and civil society the opportunity to take the lead. Partnerships with NGOs and institutions in Si Chang and Phetchaburi were crucial in bridging the gap between community priorities and scientific knowledge. Instead of being viewed as ad hoc assistance, these partnerships need to be systematically integrated into planning procedures.

Third, multifunctionality and co-benefits must be at the heart of how NbS are assessed and rewarded. All too frequently, economic productivity is viewed as a conflicting aim and biodiversity preservation and climate adaptation as distinct policy areas. This chapter's case studies demonstrate an alternative reasoning: that with careful planning, ecological restoration, livelihood resilience, and cultural continuity can all strengthen one another. A maritime zone co-managed by fishermen and divers, or a salt farm converted to biodiversity corridor, defies traditional ideas of land-use conflict and creates opportunities for collaboration. These co-benefits must be recognised and rewarded by planning systems, both through financial and regulatory frameworks.

However, the chapter also warns against romanticizing NbS as being easily transferable or generally benign. The hazards of tokenism, greenwashing, and foreign imposition are still present, especially when donor-driven objectives take precedence over local sovereignty or when NbS are viewed as flimsy additions to traditional infrastructure. Planning must adopt ethical humility to avoid these errors, recognizing that resilience is developed via sustained, place-based engagement rather than being built from above. This entails paying close attention to how historical marginalization, power imbalances, and land tenure difficulties may influence who gains and losses from adaptation initiatives.

The need for integrative, participative, and ecosystem-informed planning is growing as Thailand deals with increasing climate pressures along its coastlines, from tidal floods and saline intrusion to biodiversity loss and economic dislocation. Coastal zones are dynamic socio-ecological systems whose sustainability depends on the alignment of social justice, local governance, and ecological health. As such, they are not uniform spaces to be zoned and guarded.

In order to move forward, embeddedness must be embraced as a planning principle and a workable approach. This includes integrating NbS into local institutions, value systems, and landscapes; integrating scientific tools into community priorities; and integrating adaptation into daily life rather than just emergency response (Cheshmehzangi et al., 2024). Only with this thorough integration will NbS be able to transition from promising pilots to paradigm-shifting technologies that can successfully negotiate the challenges posed by climate change.

In conclusion, the stories of Phetchaburi and Si Chang Island are not unique achievements; rather, they are woven into a larger fabric of innovative coastal development in Thailand and elsewhere. Their dedication to listening to ecosystems, honouring local knowledge, and cooperating to create regenerative futures is what unites them. The foundation of a resilient, inclusive, and ecologically robust coastal future in Thailand will be the incorporation of these ideals into the country's coastal planning agenda.

References

Bennett, N. J., Dearden, P., Murray, G., & Kadfak, A. (2014). The capacity to adapt? Communities in a changing climate, environment, and economy on the northern Andaman coast of Thailand. *Ecology and Society, 19*(2), 5.

Bidorn, B., Phanomphongphaisarn, N., Rukvichai, C., & Kongsawadworakul, P. (2020). Evolution of mangrove muddy coast in the Western Coast of the Upper Gulf of Thailand over the past six decades. In *Estuaries and Coastal Zones in Times of Global Change: Proceedings of ICEC-2018* (pp. 429–442). Springer Singapore.

Bujang, J. S., Zakaria, M. H., & Short, F. T. (2016). Seagrass in Malaysia: Issues and challenges ahead. In *The Wetland Book*, 1–9. Springer.

Cheshmehzangi, A. (2022). Green infrastructure and urban sustainability: An editorial. In *Green Infrastructure in Chinese Cities* (pp. 1–17). Springer Nature Singapore.

Cheshmehzangi, A., Sedrez, M., & Flynn, A. (2024). Aligning with stormwater management but from the urban design perspective: Towards sustainability-driven and water-based planning and design. In *Rethinking Stormwater Management through Sustainable Urban Design* (pp. 319–327). Springer Nature Singapore.

Chirapart, A., Ohno, M., & Jarayapund, P. (2003). Marine algae of Ko Si Chang, Eastern Thailand. *Bulletin of Marine Sciences and Fisheries, Kochi University, 22*, 107–118.

Cui, L., Sun, Y., Song, M., & Zhu, L. (2020). Co-financing in the green climate fund: Lessons from the global environment facility. *Climate Policy, 20*(1), 95–108.

Fan, M., & Cheshmehzangi, A. (2022). Applying multiple nature-based solutions (NBS) with regional flexibility in bio building design in Southeast China. In *Green Infrastructure in Chinese Cities* (pp. 431–468). Springer Nature Singapore.

Kim, E. J. A., Asghar, A., & Jordan, S. (2017). A critical review of Traditional Ecological Knowledge (TEK) in science education. *Canadian Journal of Science, Mathematics and Technology Education, 17*(4), 258–270.

Kitjarak, T., Pathak, S., Qi, X., & Boripis, T. (2024). *Public service provisions for land resource planning in Khao Phanom, Krabi Province in Thailand.* https://www.preprints.org/frontend/manuscript/a4fb510fb58735a1e3da40ef19ba1f70/download_pub

Kritsanapuntu, S., Jutapruet, S., & Nakawiroj, P. (2024). Annual variations in diversity, distribution and abundance of the benthic macrofauna in the two most recreational sandy beaches on Sichang Island, Chonburi Province, in the Eastern Gulf of Thailand. *Ramkhamhaeng International Journal of Science and Technology, 7*(1), 1–15.

Li, L., Cheshmehzangi, A., Chan, F. K. S., & Ives, C. D. (2021). Mapping the research landscape of nature-based solutions in urbanism. *Sustainability, 13*(7), 3876.

Martin, J. F., Roy, E. D., Diemont, S. A., & Ferguson, B. G. (2010). Traditional Ecological Knowledge (TEK): Ideas, inspiration, and designs for ecological engineering. *Ecological Engineering, 36*(7), 839–849.

Nagabhatla, N., Hung, N. T., Tuyen, L. T. T., Cam, V. T. N., Dhanraj, J., Thien, N. T., & Swierczek, F. (2018). Ecosystem-based approach for planning research and capacity development for integrated coastal zone management in Southeast Asia. *APN Science Bulletin, 9*(1), 3–9.

Olivadese, M. (2025). Rethinking nature-based solutions: Unintended consequences, ancient wisdom, and the limits of nature. *Land, 14*(6), 1272.

Pumijumnong, N. (2013). Mangrove forests in Thailand. In *Mangrove ecosystems of Asia: Status, challenges and management strategies* (pp. 61–79). Springer New York.

Ruyabhorn, P., & Phantumvanit, D. (1988). Coastal and marine resources of Thailand: Emerging issues facing an industrializing country. *Ambio, 17*(3), 229–232.

Shoo, L. P., Catterall, C. P., Beyer, H. L., Cockbain, P., Duncan, M., Robson, T., Roche, D., Taylor, H., White, Z., & Wilson, K. (2021). Smart allocation of restoration funds over space and time. *Ecological Applications, 31*(8), Article e02448.

Tansuchat, R., & Plaiphum, S. (2023). Assessing food and livelihood security in sea salt community: A GIAHS study in Ban Laem, Phetchaburi, Thailand. *Sustainability, 15*(21), 15229.

Terton, A., Qi, J., & Jang, N. (2024). *Tracking progress on the integration of nature-based solutions and ecosystem-based adaptation in national adaptation plan processes* (NAP Global Network synthesis report). International Institute for Sustainable Development. https://napglobalnetwork.org/resource/tracking-progress-on-the-integration-of-nature-basedsolutions-and-ecosystem-based-adaptation-in-national-adaptation-plan-processes

Thiammueang, D., Juntarashote, K., & Chuenpagdee, R. (2014). Women's roles in fisheries governance in Ban Laem, Phetchaburi Province, Thailand. *IJERD–International Journal of Environmental and Rural Development, 4*(1), 45–50.

Umprasoet, W., Mu, Y., Somrup, S., Junchompoo, C., Guo, Z., & Zhang, Z. (2023). Assessment of habitat risks caused by human activities and integrated approach to marine spatial planning: The case of Sriracha District—Sichang Island. *Coasts, 3*(3), 190–208.

Vikas, N., & Hari, S. (2023). Innovative financing mechanisms to leverage ecosystem based adaptation (EbA) finance for vulnerable communities. *Global Economic Observer, 11*(1), 116–125.

Walz, Y., Nick, F., Higuera Roa, O., Nehren, U., & Sebesvari, Z. (2021). *Coherence and alignment among sustainable land management, ecosystem-based adaptation, ecosystem-based disaster risk reduction and nature-based solutions.* Institute for Environment and Human Security, UNCCD.

Chapter 6
Governance Architectures and Policy Frameworks in the Eastern Economic Corridor (EEC) Coastal Cluster: Evaluating Chonburi, Rayong, and Chachoengsao

Entry to the Chapter: Walking the Shoreline of Policy and Place
Two parallel pictures come to me when I consider the Eastern Economic Corridor (EEC), which stretches over Chonburi, Rayong, and Chachoengsao. One is the well-polished image of the EEC as a center of industrial efficiency, a trade entry point, and an investment magnet that is displayed in glossy brochures and government initiatives. The other is the reality I have witnessed on the ground, in the streams that bring pollution and life to the sea, in the uneven faces of communities negotiating fast change, and in the air heavy with industrial heat. It is a place where government is a dynamic, contested area rather than an abstract system, shaped by the interaction of environmental vulnerability, economic ambition, and the tenacity of those who live along this corridor.

In Chonburi, I walked stretches of the coast where the sea meets vast industrial estates. Located amid a fragile coastal system where mangroves originally protected the beach and absorbed storm shocks, these estates are the pride of the country's economic planning. In place of concrete infrastructure that reflects waves rather than absorbs them, many of those natural buffers have either been reduced in size or eliminated. Talks with locals revealed a mixture of insecurity as they mentioned frequent flooding, rising tides, and the gradual infiltration of saltwater into fields, as well as pride in the job prospects, better infrastructure, and increasing property values. Here, governance frequently appears to be torn between addressing the nagging dangers that could threaten the EEC's basic foundation and advancing its high-tech, export-driven image.

Rayong's shoreline told a similar story, though in a different dialect. This province has a long history of heavy industry and petrochemicals, and industrial policy and government are closely related. The horizon is dominated by industrial ports and refineries, yet beyond them are smallholder farms and fishing villages that have survived for many generations. During my visits, I observed that environmental regulations are frequently reactive rather than integrated into long-term frameworks for climate resilience, such as tidal floods that lead to temporary drainage solutions

A. Cheshmehzangi, *Thailand's Tropical Thresholds*, Urban Sustainability,
https://doi.org/10.1007/978-981-95-4975-7_6

or industrial spills that result in rushed cleanup efforts. I could feel the strain on governance structures that are responsible for safeguarding the ecological integrity of a coast that is being put to the test more and more by heat waves, torrential rains, and changing monsoon patterns, as well as a globally competitive industrial base.

Chachoengsao, the quieter of the three provinces in the EEC coastal cluster, surprised me with its layers of vulnerability. Its relationship to the EEC is more about logistics, agriculture, and land conversion than it is about ports, as it is partially located along the Bang Pakong River and inland floodplains. Here, zoning, riverbank protection, and land-use modification decisions made by the government have an impact downstream, both physically and figuratively, into the center of the corridor. I stood on embankments built to hold back river floods and wondered how they would fare in a decade, when sea level rise forces saltwater farther inland and upstream rainfall increases. Though the local realities—subsiding land, changed water flows, and the deterioration of community-led flood management traditions—call for a more grounded, place-specific approach, the policies influencing Chachoengsao's future frequently take inspiration from national development models.

I observed the EEC coastal cluster's bold yet fragile government in all three provinces. Setting goals for infrastructural improvements, foreign investment, and economic transformation is bold. It is weak in its ability to incorporate climate adaptation into these same goals. The governance architectures I saw frequently function in silos: environmental agencies are issuing compliance guidelines, provincial authorities are juggling political commitments, economic ministries are striving for expansion, and local administrations are fighting for a voice in decisions that will change their landscapes. The problem is not only a lack of policies; it is also the way rules are layered and do not always correspond, leaving gaps that allow climate hazards to go unreported or neglected.

I found myself thinking about the corridor not only as a physical space, but as a governance experiment in real time. Despite being implemented in one of Thailand's most climate-exposed locations, the EEC is intended to hasten Thailand's entry into global value chains. Highways cut across agricultural and coastal areas, ports are deepened to handle larger ships, and industrial estates grow on reclaimed land. There are climate costs associated with each intervention, such as decreased ecological buffers, changed drainage patterns, and more impermeable surfaces. The policies that facilitate this expansion don't always recognize how these effects compound, and they don't always call for adaptation strategies commensurate with the level of development.

I was really surprised by how unequal the involvement was. The most vulnerable groups, such as small-scale fishermen, subsistence farmers, and residents of informal settlements, were frequently either absent or present but mute at meetings and public discussions, despite the formal institutions of stakeholder engagement being in place. In the meantime, regional elites and business players could influence policy timelines and narratives to suit their interests. This imbalance results in a governance architecture that is both technically advanced and socially vulnerable. It can swiftly complete massive infrastructure projects, but it takes longer to address the long-term and diffuse dangers posed by climate change.

But there were times when it seemed like there may be another way. I met a group of community leaders in Chonburi who were promoting the restoration of mangroves as a flood control measure. They presented this idea as a type of protective infrastructure rather than an environmental initiative. To identify heat stress zones and suggest changes to the urban architecture of industrial towns, local scholars collaborated with provincial officials in Rayong. I learned about farmer cooperatives in Chachoengsao testing saline-resistant crops, which are an adaptation to the encroaching salinity that legislative frameworks have not yet fully recognized. These instances made me realize that governance is not just a matter that takes place in boardrooms and ministries; it is also negotiated on a daily basis in the marketplaces, the fields, and the neighbourhoods that are most vulnerable.

As I left the EEC coastal cluster, I carried a sense of urgency. The frameworks for governance and policy in this area are at a crossroads: they can either keep putting economic acceleration ahead of climate resilience or they may shift to an integrated model that views adaptation as an integral part of development. Based on what I've witnessed and heard, I'm worried that if this change doesn't happen, the industries and infrastructure in the corridor would eventually be on shaky ground, both literally and figuratively, as they are undermined by the very environmental factors for which they weren't prepared.

The Eastern Economic Corridor is not just a symbol of Thailand's economic future; it is also a test of whether governance can be both visionary and adaptive. My hope is that the lessons from its coasts, rivers, and communities will be taken seriously—before the next decade's photographs tell a story not of progress secured, but of opportunities missed.

6.1 Introduction: Governing for Climate Resilience in Coastal Thailand

Thailand's coastal regions are now frontline areas where environmental instability and economic aspirations conflict as climate hazards increase throughout Southeast Asia. The socio-ecological features of Thailand's eastern seaboard are already changing due to storm surges, sea level rise, salt intrusion, and rising heat extremes. Furthermore, it is becoming increasingly evident that institutional capacity—laws, plans, policies, and governance arrangements—is just as important in deciding whether coastal towns adapt or fail, even if ecological fragility frequently takes center stage in climate debate. In this regard, governance frameworks and ecosystems play an equal role in climate resilience.

This is particularly noticeable in Thailand's premier regional development project, the Eastern Economic Corridor (EEC), which includes the provinces of Chonburi, Rayong, and Chachoengsao (Tontisirin et al., 2017). The EEC was introduced in 2017 as a component of Thailand 4.0, a national strategy to move toward an economy centered on innovation and knowledge (Kohpaiboon, 2020). It is envisioned as a high-tech industrial and logistical hub that is connected to international supply chains and equipped with smart city technology. Beneath this image of rapid economic expansion, however, is a complicated governance challenge: how to balance local ecological restrictions, institutional variety, and uneven planning capacities with national development aspirations.

Because of its unique location, the EEC coastal cluster can be used as a testbed to better study Thailand's climate governance dynamics (Bhrammanachote, 2019). It provides a multi-layered governance environment where transboundary environmental issues, local government mandates, province planning regimes, and national economic policies all come together. Despite having different political cultures, natural settings, and socioeconomic systems, these three provinces are being brought together under a single legislative and administrative framework to pursue a shared growth path. The EEC Act of 2018 gives national authorities broad authority to supersede regional spatial plans and accelerate infrastructure spending (Secchi & Belladonna, 2020). At a time when climate adaptation necessitates local knowledge, stakeholder inclusion, and context-specific responses, this might potentially upset long-standing ecological practices and marginalize local voices, even while it might provide procedural efficiency.

This chapter uses the EEC as a strategic case study to analyze climate resilience governance in Thailand and assess how institutional frameworks either facilitate or impede effective adaptation. It accomplishes this by examining three fundamental aspects: (1) the coherence and coordination of national and local policy frameworks; (2) the influence of decentralization on subnational climate planning capabilities; and (3) the contributions of non-state actors, including universities, civil society organizations, and private sector stakeholders, on the development and implementation of climate action.

Thailand has made significant progress in integrating climate resilience into its national policy framework throughout the last ten years. Five major sectors—water management, agriculture, health, tourism, and natural resources—are listed as adaptation priorities in the 2018 National Adaptation Plan (NAP) (Kiguchi et al., 2021). Additionally, it promotes incorporating adaptability into the planning of local growth (Meeros & Wichakul, 2022). In the meantime, a long-term vision for a low-carbon, climate-resilient economy is provided by the Climate Change Master Plan (2015–2050) (Tongdethsri, 2021). These plans, however, work in tandem with other national frameworks that treat land use, coastal protection, and environmental safeguards differently, such as Thailand's Spatial Development Master Plan, the EEC Development Plan, and the 12th National Economic and Social Development Plan.

This policy fragmentation is most noticeable in areas that are changing quickly, such as the EEC, where sensitive coastline geographies are overlaid with transportation megaprojects, industrial corridors, and smart city plans. The EEC provinces are already dealing with a number of climate-related stressors: Chachoengsao is seeing land use conflicts between its agro-ecological identity and encroaching industrial zoning (Molle & Srijantr, 2003); Rayong is dealing with hazardous waste and estuarine degradation as a result of its industrial estates; and Chonburi is facing coastal erosion, flooding, and marine pollution from port expansion. A variety of urban, peri-urban, and rural settings can be found in each province, resulting in intricate governance levels that must strike a balance between the demands of economic growth and adaptation.

The decentralization reforms that have been in place since the late 1990s have made matters more challenging. Local governments, including municipalities and Tambon (sub-district) administrative organizations, were given more administrative and financial power as a result of these reforms (Kamnuansilpa, 2014). Theoretically, this paradigm allows for locally driven climate planning that is adapted to community needs and particular vulnerabilities. Decentralization has, however, been unequal in practice and frequently undermined by erratic policy directives, a lack of technical assistance, and restricted fiscal autonomy. This is particularly true in development zones that receive a lot of attention, like the EEC, where local autonomy is usually subordinated to national interests. Because of this, the EEC provides a rich environment for investigating the relationship between decentralization and climate governance in situations where local authority is both empowered and limited.

Non-state actors are increasingly influencing the agenda for climate resilience within this changing governance matrix. Research and community involvement initiatives on coastal vulnerability, ecological restoration, and adaptive land use have been spearheaded by universities including Burapha University and Kasetsart University. In order to protect mangrove habitats, promote environmental justice, and strengthen local adaptability, civil society organizations have organized. Meanwhile, there is growing pressure on private sector actors to implement risk management, green certification, and environmental norms, especially those running industrial estates and logistics hubs. These varied actors frequently function alongside official governmental entities, opening chances for cooperation while also highlighting inconsistencies in responsibility and coordination.

Therefore, it is necessary to go beyond institutional checklists or policy inventories to comprehend the governance of climate resilience in the EEC. It necessitates a relational study of the movement of resources, power, and knowledge across sectors and scales. Additionally, it calls for consideration of governance processes' implementation as well as their design. Who is involved in making decisions? Whose expertise is it? What systems are in place for learning, feedback, and course correction? In order to determine what facilitates and hinders adaptation in intricate, multi-level governance systems, these questions are crucial.

These dynamics are then broken down into five interconnected sections in this chapter. The EEC governance regime's national-local policy interactions are examined in Sect. 6.2, which also highlights potential and conflicts in coordinating adaptation with industrial development. The role of non-state actors is examined in Sect. 6.3, which evaluates how advocacy networks and knowledge collaborations either support or impede climate action. The ability of local governments to develop and carry out adaptation strategies is evaluated in Sect. 6.4, with an emphasis on both institutional limitations and new developments. In conclusion, Sect. 6.5 summarizes the most important findings and suggests an integrated, adaptive governance model that strikes a compromise between Thailand's national goals and the demands of local resilience.

Finally, efforts to control climate resilience in Thailand's coastal regions are still in the early stages of development. Chonburi, Rayong, and Chachoengsao, which make up the EEC coastal cluster, act as a living laboratory that demonstrates the potential and drawbacks of integrated climate governance. By using this perspective, the chapter not only advances our knowledge of Thailand's adaptation environment but also provides more general insights into how the Global South's rapidly growing coastal regions can deal with the climate crisis without compromising ecological integrity or democratic accountability.

6.2 National-Local Dynamics in the EEC: Tensions, Gaps, and Overlaps

National policy frameworks and municipal planning systems in Thailand's Eastern Economic Corridor (EEC) are becoming increasingly involved in a delicate debate between climate vulnerability and economic ambition (Jaiprasert & Maluleem, 2023). Although the EEC is intended to expedite Thailand's shift to a contemporary industrial economy, its institutional and legislative framework highlights underlying conflicts between decentralized governance and centralized control, especially with regard to the incorporation—or lack thereof—of environmental planning and climate adaptation into provincial and local development decisions.

The EEC Act from 2018, which gives national organizations like the National Committee for the Eastern Special Development Zone and the EEC Office broad jurisdiction (Tontisirin & Anantsuksomsri, 2021), is at the heart of this conflict. Through unique regulatory procedures, these organizations can accelerate environmental impact evaluations, supersede provincial land-use plans, and accelerate infrastructure projects. Although the goal of these powers is to expedite development, they also create conflicts with Thailand's current decentralization system, which supposedly gives subnational governments the authority to create and carry out local environmental and development plans.

Chachoengsao, a province renowned for its agricultural landscapes, water-dependent ecosystems, and long-standing indigenous resistance to unbridled industrialization, is particularly affected by this governance mismatch. Here, provincial conservation concerns have battled with the top-down imposition of EEC industrial zones and logistics hubs, particularly in the Bang Pakong River Basin and flood-prone farming districts. The province's ecological reality and climatic exposure are not reflected in national land-use regulations, according to local officials. Furthermore, communities and civil society organizations contend that environmental impact evaluations are carried out after the fact or without significant public involvement, which exacerbates the mistrust that exists between local constituencies and central planners.

The scenario in Chonburi illustrates a distinct yet no less intricate intersection of local and national concerns. Chonburi has had substantial public and private investment in logistics and infrastructure since it is the location of Laem Chabang Port and other key EEC investment zones (Bhrammanachote, 2019). However, urban flooding, marine pollution, and coastal erosion have all been exacerbated by this rapid expansion. The Provincial Administrative Organization of Chonburi has made an effort to incorporate climate resilience into its strategic plan, which includes plans for ecosystem restoration and green infrastructure. However, it is still hindered by overlapping responsibilities and its limited capacity to change national spatial plans. In this case, the province lacks the means to negotiate climate-compatible alternatives inside the strict EEC framework, therefore the gap is not resistance but structural incompetence.

Another variant is provided by Rayong. The most industrialized province in Thailand is dealing with growing health and environmental issues related to hazardous waste and petrochemical development. Provincial officials are under growing public pressure to enact stronger environmental regulations and prepare for climate hazards, even if they have shifted their priorities more in line with the national economy. Managing the confluence of industrial growth, environmental degradation, and community resistance under inadequate coordination systems is Rayong's challenge (Kim & Aroonpipat, 2016), not legal exclusion.

A recurring theme in all three provinces is the discrepancy between the spatial planning tools used under the EEC and the National Adaptation Plan (NAP). In the EEC setting, the NAP has little bearing on actual land distribution, infrastructure investment, or environmental permitting, despite its demands for mainstreaming adaptation into sectoral and municipal development plans (Ngernthuam et al., 2024). Furthermore, the comprehensive, systems-oriented logic of climate resilience is undermined by Thailand's sector-based approach to policy implementation, in which ministries supervise discrete duties. The ministries of agriculture, industry, natural resources, and interior each have their own priorities, frequently without unified frameworks for land-use planning, disaster risk reduction, and adaptation.

Institutional gaps in data sharing, policy enforcement, and decision-making result from this fragmented governance environment. Because they are expected to carry out centrally planned programs with limited leeway or financial autonomy, local governments frequently find themselves in reactive roles. In the meanwhile, there are no reliable ways for national authorities to integrate traditional ecological methods, civil society input, or local climate risk knowledge into project design.

In conclusion, the EEC serves as an example of how conflicts between local and national government institutions can impede the development of effective plans for climate resilience. Climate action is marginalized—treated as a compliance issue rather as a fundamental concern—when economic development is given precedence over all other considerations and decentralized institutions lack actual authority or integration channels. More than just harmonizing policy texts is needed to close this gap; fundamental reform in the way authority, accountability, and information are distributed within Thailand's multi-level governance structure is required.

6.2.1 *The EEC Act and Its Implications for Provincial Autonomy*

A major turning point in Thailand's national development plan was the passage of the Eastern Economic Corridor Act B.E. 2561 in 2018, which established the legal framework for an ambitious regional transformation along the eastern seaboard of the country (Niyomsilp et al., 2020). The EEC Act gives central authorities extraordinary powers in an effort to draw in foreign investment, expedite infrastructure improvements, and establish Thailand as a regional center for innovation and logistics. Although the Act is successful in expediting approvals and providing incentives for rapid economic growth, it raises serious concerns about the loss of provincial autonomy, the deterioration of democratic planning procedures, and the exclusion of environmental and climate-related factors from subnational governance.

Fundamentally, the provinces of Chonburi, Rayong, and Chachoengsao are included in the special development zone established by the EEC Act. The EEC Policy Committee, a national-level authority presided over by the prime minister and advised by top economic and technological experts, regulates land use, investment licensing, and environmental review processes in this zone rather than traditional provincial or municipal legislation. The EEC Office can reinterpret land uses, expedite approvals for large-scale infrastructure, and engage in direct negotiations with private investors thanks to this centralized design, which successfully circumvents provincial planning procedures.

As a result, provincial administrations are unable to influence the destiny of their own regions due to a highly vertical government style. Municipal governments and Provincial Administrative Organizations (PAOs) are still in charge of basic land management, local infrastructure upkeep, and service delivery, but they are mostly left out of decisions about industrial zoning, mega-project alignment, and coastal transformation (Jongtong, 2021; Nuamcharoen & Sattakorn, 2021). The overarching EEC Strategic Plan, which frequently imposes preset economic zones, transport corridors, and utility corridors, supersedes subnational plans, even if they exist. Examples of these plans are Provincial Spatial Development Plans and Municipal Development Strategies.

There are significant ramifications for the governance of climate resilience. For instance, the EEC framework is being used in Chonburi to expand the Laem Chabang Port, with no consideration given to local concerns about disaster readiness, urban heat, or water stress (Teerawattana & Yang, 2019). Although the municipality has created climate-relevant strategies, such as stormwater management zones and green urban corridors, these are ancillary to the main EEC expenditures that are changing the shoreline. This is also the case in Rayong, where the EEC's focus on smart logistics and petrochemical clusters has superseded established ecological rehabilitation plans in mangrove and damaged estuary areas. Even when local officials voice concerns, they sometimes lack the institutional power or regulatory clout to force changes to the designs of EEC projects.

Perhaps the most contested case is that of Chachoengsao, where local governments and communities have been more openly resistant to EEC encroachment. Chachoengsao has long preserved a semi-rural character focused on agriculture, conservation, and agrotourism. It is well-known for its natural richness, flood-prone lowlands, and cultural landscapes along the Bang Pakong River (Wongsa et al., 2020). Large areas of land, however, are being repurposed for inland ports, logistical depots, and industrial estates under the EEC plan. Officials from the province have openly voiced their worry that these changes will make it more difficult for them to control flood risks, preserve wetland ecosystems, and uphold the integrity of the regional agricultural sector. However, they only have a few intervention tools: Public hearings are frequently performed late in the process with predetermined results, land expropriations are expedited, and Environmental Impact Assessments (EIAs) are carried out through national authorities.

As a result, the EEC Act establishes a systemic imbalance between provincial climate resilience and national economic planning. Although the goal of Thailand's decentralization laws, especially the 1999 Decentralization Act, was to give subnational governments more authority over finances, administration, and planning, the EEC is a major step back from those goals (Suwanmala & Weist, 2009). It gives actors who are primarily responsible for meeting national economic goals—rather than local ecological realities or community well-being—control over climate-sensitive land use decisions.

Furthermore, rather than strengthening environmental protections, the Act makes it easier to streamline environmental governance. For instance, the EEC law does not require Strategic Environmental Assessments (SEAs), and Environmental Impact Assessments are frequently expedited for "priority projects," which reduces the time and space available for critical examination. It is uncommon to take into account the combined effects of several developments, especially in climate-sensitive regions like coastal zones, flood plains, or aquifer recharge areas. This regulatory deficiency is more than just a technical error; it reflects a larger institutional logic that prioritizes economic needs over environmental and climate concerns.

From a policy standpoint, the EEC Act is an example of technocratic centralism, wherein elite decision-makers are trusted to manage complexity through investor confidence mechanisms, smart zoning, and streamlined processes. The flexible, polycentric, and participatory types of governance that climate resilience demands are incompatible with this, even though it can be justified in situations of national disaster or institutional paralysis. Because climate hazards are localized, unpredictable, and dynamic, governance frameworks must be adaptable, inclusive, and able to absorb information from a variety of sources.

This is not to say that climate-responsive innovation is not welcome in the EEC. Municipalities in Chonburi and Rayong are experimenting with risk-sensitive infrastructure, urban greening, and adaptation education for the community. In order to investigate low-carbon industrial estates and green port technologies, the EEC Office has also collaborated with academic institutions and think tanks. However, because the EEC's fundamental governance logic places a higher priority on economic acceleration than ecological integration, these initiatives continue to be ancillary, i.e., pilots rather than paradigms.

In conclusion, the EEC Act is a crucial test of Thailand's ability to strike a balance between climate resilience and economic change. Its current design marginalizes environmental control, circumvents local adaptation planning, and threatens provincial autonomy. Recalibrating the EEC to strengthen provincial institutions, restore the integrity of environmental assessments, and institutionalize climate-sensitive planning as a shared responsibility across governance levels is necessary if it is to be an example of sustainable development rather than just industrial growth. The EEC runs the risk of turning into a corridor of increasing vulnerability rather than one of resilience in the absence of such reforms.

6.2.2 Incoherence in Planning Mandates: Chachoengsao's Resistance and Reinvention

Chachoengsao is the most active site of resistance and alternative planning among the three provinces that make up the Eastern Economic Corridor (EEC) (Bhrammanachote, 2019). Because of its vast wetlands, rice fields, and cultural landscapes along the Bang Pakong River, Chachoengsao has managed to preserve a strong eco-agricultural identity, in contrast to Chonburi and Rayong, where industrialization has deeper historical roots. Traditional farming methods, seasonal flooding, and water-based livelihoods are all intricately woven into its socio-ecological fabric (Bhrammanachote, 2019). However, under the EEC framework, large tracts of this fragile terrain are being set aside for inland ports, logistical centers, and industrial estates, resulting in stark conflicts between local ecological goals and national development mandates.

A profound incoherence in Thailand's multi-level planning framework has been revealed by this friction. On paper, the provincial development plan for Chachoengsao places a strong emphasis on flood control, agrotourism, and environmental preservation. However, central agencies empowered by the EEC Act are progressively overriding these objectives by reclassifying land for industrial use without adequately integrating it with local spatial policies or climate risk assessments. As a result, there is a planning dilemma whereby municipal officials are required to support projects that jeopardize ecological stability and hydrological function while also implementing climate resilience measures.

Chachoengsao has become a symbol of institutional transformation in response. Religious leaders, civil society organizations, and local governments have banded together to openly criticize some EEC developments. Using ecological zoning, cultural heritage protections, and participatory land use mapping, alternative visions have been put forth, such as "Green Chachoengsao" and "Agricultural Heritage Corridors" (Charoentrakulpeeti et al., 2021). In an effort to regain some technical control over land distribution and climate planning, the province government has also teamed up with academic institutions to carry out Strategic Environmental Assessments (SEAs) separate from national EIA procedures.

This case demonstrates both the inventive methods in which provincial actors get around institutional restraints and the boundaries of local authority under centralized economic regimes. Chachoengsao's opposition represents an emerging paradigm of climate-conscious planning from below, which questions the developmental orthodoxy of the EEC and reasserts the significance of ecological identity in determining future paths, even while legal authority is still biased toward national agencies.

6.3 The Role of Non-state Actors: Universities, NGOs, and Private Sector Engagement

Non-state actors have a growing impact on how climate resilience and environmental stewardship are shaped in the Eastern Economic Corridor's (EEC) governance environment. Universities, non-governmental organizations (NGOs), and private sector actors have entered crucial arenas for action, negotiation, and innovation as national policy frameworks and local government institutions attempt to balance ecological stability with growing industrialization. Even though they are frequently dispersed and function outside of established power structures, these players have played a significant role in promoting climate governance in the provinces of Rayong, Chachoengsao, and Chonburi—albeit with varying degrees of success.

Universities are becoming important knowledge centers that connect science, community involvement, and policy recommendations. Organizations like Kasetsart University's Sriracha Campus and Burapha University in Chonburi have established comprehensive research programs on flood risk mapping, mangrove restoration, coastal ecology, and climate adaptation planning. For instance, Burapha University has collaborated extensively with Chonburi's local governments to implement green infrastructure techniques like bioswale construction, permeable surface design, and coastal buffer restoration. These programs frequently have two functions: they act as real-world demonstration projects, and they provide forums for knowledge co-production among communities, local authorities, and researchers.

Kasetsart University in Rayong has taken the lead in evaluating climate vulnerabilities in industrial areas, with an emphasis on flood exposure, heat island effects, and air quality. The university has been instrumental in bringing together private businesses, local governments, and the Industrial Estate Authority of Thailand (IEAT) to investigate adaptation strategies such low-carbon logistics, rooftop solar, and wastewater recycling (Yamsrual & Potipituk, 2021). Universities are producing useful data and scenario modelling, but their impact on real regulatory procedures is still limited. Much of this work runs the risk of being advisory rather than revolutionary in the absence of official avenues to incorporate scholarly results into policy.

In Chachoengsao, where grassroots mobilization has been essential to opposing EEC-driven industrial expansion, NGOs have adopted a more combative and community-focused stance. To replant mangroves, document environmental impacts, and engage in participatory land use mapping, organizations such as the Eastern People's Network for Environmental Justice have collaborated with local temples, farmers, and youth groups. These initiatives are based on long-standing customs of community rights and environmental action, frequently referencing Indigenous land knowledge and Buddhist ecological ethics.

The "Living Bang Pakong" campaign is one noteworthy effort that portrays the river as a cultural and spiritual lifeline in addition to an ecological treasure (Chaiyarak et al., 2019). To challenge official evaluations and sway public opinion, NGOs in this area have partnered with researchers to generate citizen science data on biodiversity loss and water quality. NGOs frequently lack steady funding, legal status in planning forums, and the power to halt state-approved projects, despite having significant moral authority. Their success depends on forming partnerships with media organizations, empathetic academics, and occasionally local political players who are prepared to oppose centralized orders.

In contrast, the private sector is a more conflicted but growing force in the field of climate governance. Some businesses in Rayong's Map Ta Phut Industrial Estate have embraced voluntary sustainability norms, particularly those aiming for international accreditation or ESG (Environmental, Social, and Governance) alignment. These include the application of pollution monitoring systems, circular economy ideas, and green building requirements. Global companies in the automotive and chemical industries have also started corporate social responsibility (CSR) initiatives centered on climate education, community replanting, and coastal cleanup.

One such instance is the establishment of a "Climate Learning Garden" on the estate's border through a public–private cooperation between a nearby school and the IEAT, a prominent petrochemical company (IEAT, n.d.). In order to increase awareness among young people and staff, the project includes solar panels, aquaponic systems, and climate change exhibits. Although such efforts are admirable, they are nonetheless optional, uncontrolled, and frequently unrelated to systemic change. Furthermore, these initiatives don't often address the main environmental threats that come from industrial processes themselves, like resource overextraction, air pollution, and chemical leaks.

These instances demonstrate the increasing importance of non-state players in climate policy, but they also highlight their dispersed authority. Private entities provide funding and technical skills, NGOs mobilize local knowledge and public accountability, while universities give science and credibility. However, their combined impact is restricted in the lack of a formal governance framework that incorporates these contributions. Promising experiments frequently stay isolated or are stopped once project funding expires due to a lack of coordination methods, feedback loops, and shared monitoring systems.

Non-state players would be acknowledged as co-governors of climate resilience rather than as ancillary stakeholders in a more integrated governance model. Formalizing their involvement in EEC planning procedures, developing hybrid policy platforms, and providing institutional incentives for cooperative experimentation would all be necessary to achieve this. For example, academic researchers, representatives of civil society, and commercial sector actors could be appointed as permanent members of provincial adaptation councils, with specific responsibilities in determining the agenda and making decisions.

In conclusion, the EEC cluster serves as an example of the potential and risk associated with the involvement of non-state actors in climate governance. The majority of innovation and moral clarity can be found in the interstitial spaces, such as university partnerships, NGO campaigns, and CSR pilot initiatives, even while the formal governing machinery is still primarily top-down and driven by the market. Deliberate institutional design, trust-building, and a change in the way legitimacy is conceptualized within Thailand's development paradigm are all necessary to unlock the revolutionary potential of these players.

6.4 Local Government Capacity and Decentralised Adaptation Planning

The goal of Thailand's decentralization reforms, which were started in the late 1990s, was to provide subnational governments more financial resources, administrative autonomy, and planning power so they could better meet the requirements of local development. Theoretically, these changes made local governments—in particular, municipalities, Tambon Administrative Organizations (TAOs), and Provincial Administrative Organizations (PAOs)—essential players in the planning of climate adaptation. However, due to institutional dependencies and structural constraints, local governments' ability to execute effective climate resilience plans is still quite inconsistent in practice, particularly in the rapidly rising Eastern Economic Corridor (EEC).

The capacity of local governments to address climate concerns varies greatly between the EEC provinces of Chonburi, Rayong, and Chachoengsao. By incorporating urban heat mitigation techniques, drainage improvements, and green infrastructure into local planning documents, some municipalities—like Chonburi's Laem Chabang City Municipality—have shown a great deal of leadership. In addition to being close to educational institutions like Burapha University that promote technical capacity-building, they enjoy a comparatively large revenue base as a result of port and industrial activity.

These instances are the rare, though. Adaptive capacity is severely constrained for most municipalities and TAOs in the EEC by limited planning autonomy, staff shortages, and funding constraints. Basic service delivery and infrastructure upkeep frequently take up annual budgets, leaving little space for environmental restoration or climate innovation. Furthermore, most local authorities lack the institutional capacity to manage the planning challenges brought about by the complexity of EEC-induced development, especially when it comes to cross-sectoral climate hazards that affect energy, health, land use, and water.

The lack of integrated, climate-informed planning tools at the municipal level is a major obstacle. There is no legal mandate or technical framework to operationalize adaptation at subnational levels, despite Thailand's National Adaptation Plan encouraging local governments to include it in their development plans. Because of this, local plans for climate adaptation are still primarily aspirational, lacking monitoring indicators, ambiguous implementation pathways, and little baseline data. Many EEC municipalities still rely more on investor-driven zoning to influence their spatial plans than on heat exposure or flood risk mapping.

Moreover, there is still a lack of institutional coordination. Theoretically, decentralization encourages cross-departmental integration, but in reality, local departments in charge of environmental protection, public works, and disaster prevention frequently work independently. When national ministries maintain ultimate control over industrial and infrastructural projects, interagency cooperation is especially weak, limiting local governments to the function of service providers or facilitators rather than strategic planners.

Another issue is the inclusion of stakeholders. Public consultations are a common aspect of the planning process in many municipalities, however they are sometimes superficial and not truly participatory (Cheshmehzangi, 2021; Cheshmehzangi & Dawodu, 2018). Residents of regions undergoing fast EEC-driven change, like Rayong's peri-urban districts or Chachoengsao's agricultural outskirts, commonly complain that their worries about water scarcity or environmental degradation are disregarded in favour of economic development demands (Teo et al., 2019; Yu et al., 2025). When communities feel left out of choices that alter their landscapes and means of subsistence (Cheshmehzangi, 2025), trust in local institutions declines.

However, indications of institutional innovation are beginning to appear. A few local governments are experimenting with nature-based drainage systems, community-based disaster planning, and participatory budgeting. Donor-supported pilot programs have introduced training modules on climate risk assessment and urban resilience, often in partnership with Thai universities. Despite their modest size, these programs show that local governments can serve as catalysts for adaptive planning that is suited to the demands of particular locations if given the proper assistance.

In conclusion, Thailand's decentralized climate governance holds great promise, but it is now falling short of expectations in the context of the EEC. The ability of municipalities and TAOs to take the lead on adaptation has been hampered by the prevalence of top-down EEC planning frameworks, inadequate institutional capacity, and restricted local fiscal space. In addition to more funds, strengthening subnational governance for climate resilience would necessitate giving local governments the institutional authority to jointly develop, carry out, and oversee climate policies with communities and non-state actors.

6.4.1 Capacity Gaps and Planning Constraints in Secondary Cities: The Case of Rayong

Under the Eastern Economic Corridor (EEC) development framework, Rayong, one of Thailand's most industrialized provinces, has led the way in economic growth. The province has benefited from significant petrochemical investments, manufacturing clusters, and infrastructure projects, but its secondary cities and smaller municipalities have had to contend with a distinct set of difficulties as they are torn between the constraints of inadequate planning systems and the demands of rapidly increasing urbanization. These regions, which are frequently located between peri-urban communities and industrial zones, are becoming more susceptible to climate hazards like heat stress, flooding, and pollution of the air and water. Yet, a number of operational and structural shortcomings limit their ability to react.

The vulnerability of local governments' budgets is one of the main problems. While Rayong's subsidiary municipalities, like Ban Chang, Pluak Daeng, and Nikhom Phatthana, have much smaller budgets, the city's center metropolitan districts have greater revenue streams from taxes, industry contributions, and tourism (Srisomyong, 2010). These are frequently overburdened to provide public services, garbage management, sanitation, and basic infrastructure. Unless it is connected to external donor programs or one-time government subsidies, funding for environmental restoration, climate adaption, or green infrastructure is usually nonexistent. As a result, even though environmental concerns are becoming more urgent, climate resilience planning is still a low priority issue on municipal agendas.

There is a large technical capability gap in addition to financial limitations. Access to qualified urban planners, environmental engineers, or data analysts with experience in climate risk management is limited in many of Rayong's minor communities. Many municipal employees are generalists who handle a variety of tasks, such as public works and land management, which leaves little time for comprehensive climate planning. Furthermore, national agencies' capacity-building initiatives are few and mostly theoretical, offering few useful resources for incorporating adaptation into regional strategies.

The lack of climate-relevant data and decision-support tools is a particularly serious issue. Although academic establishments like Kasetsart University have created flood models and vulnerability assessments for Rayong province, these are not consistently included into municipal planning procedures. Anecdotal risk knowledge, paper-based land use plans, and antiquated maps are frequently used by local authorities. The ability to analyze satellite data, monitor hazards in real time, and create future scenarios is limited. As a result, a lot of towns still permit construction in flood-prone locations or disregard zoning laws that would lessen long-term exposure to environmental risks.

Another obstacle is the institutional framework of planning itself. Local development plans are frequently reactive, concentrating on meeting short-term growth demands like housing demand, road construction, and investor interest rather than long-term planning for climate adaption or environmental preservation. Documents related to spatial planning are frequently disjointed from other tools, such as plans for disaster risk reduction, health monitoring, or water resource management. Even when there are national or provincial climate adaptation guidelines, they are rarely operationalized or localized at the local level.

The absence of mechanisms for participation exacerbates these issues. In Rayong, a lot of secondary cities only hold formal hearings with little opportunity for in-depth public comment on planning issues. Communities that are directly affected by climate change and whose local knowledge could guide adaptive measures are alienated by this. Residents in low-lying and coastal locations, for example, frequently identify early indicators of saline intrusion, subsidence, or microclimatic alterations; yet planning choices hardly ever take these observations into account.

There are areas of innovation in spite of these limitations. For instance, in Ban Chang Municipality, a small group of local leaders has joined forces with a project sponsored by a university to create a green corridor that serves as a flood barrier and a recreational area by connecting urban parks to drainage canals. To promote civic consciousness, local schools in Pluak Daeng have incorporated climate education and community tree planting into their environmental curricula. Despite their modest size, these initiatives show how local players may innovate if given the proper encouragement and institutional flexibility.

In conclusion, the situation of Rayong's secondary municipalities demonstrates the conflict between the need for development and the ability to plan for climate change. There is a chance that adaptation may continue to be an afterthought as these cities grow beneath the EEC's shadow, addressed exclusively during emergencies rather than via proactive governance. It will take consistent investment in municipal financing, planning knowledge, data systems, and inclusive governance practices that prioritize climate resilience in local development strategy to close this gap.

6.4.2 Institutional Innovation at the Margins: Laem Chabang Municipality in Chonburi

Laem Chabang's initiatives to mainstream climate resilience through early warning systems, adaptive zoning tools, and green infrastructure projects are highlighted in this subsection. It demonstrates how structural limitations can be somewhat mitigated by institutional leadership and city-university collaborations. Even though the Eastern Economic Corridor (EEC) is dominated by top-down planning, Chonburi's Laem Chabang Municipality provides a striking illustration of institutional innovation from inside the local governance structure. Laem Chabang, Thailand's largest port city and a vital hub in the EEC's industrial and logistical network, is subject to severe environmental stresses, including urban heat islands, flood hazards, air and water pollution, and coastline erosion. Municipal officials, however, have deliberately created plans to incorporate climate resilience into urban development rather than passively following national directions.

Implementing green infrastructure projects, such as vegetated detention ponds, roadside bioswales, and permeable pavements in new development zones, has been a key component of Laem Chabang's strategy. Localized flood mapping, carried out in collaboration with Burapha University, which supplied both technical know-how and monitoring equipment, served as the basis for these actions. In order to decrease runoff and partially restore wetland functions in peri-urban regions, the city has also made investments to improve its stormwater management systems and incorporate biological corridors into drainage networks.

A real-time early warning system for flash floods and coastal surges has also been introduced by Laem Chabang. Rainfall thresholds, automatic weather sensors, and digital notifications distributed via SMS and municipal applications are all part of this system. Even while its scope is currently constrained, it represents a move toward proactive, data-driven governance, especially in low-lying residential areas and industrial zones with elevated disaster risk. With assistance from regional disaster risk reduction platforms, municipal employees have received training on emergency response coordination and scenario planning.

Most notably, Laem Chabang has been experimenting with techniques for adaptive zoning. In order to restrict high-density development and encourage buffer planting and community space, the municipality worked with provincial authorities and EEC planners to categorize several coastal and flood-prone regions as mixed-use green transition zones. These zoning adjustments, though modest, reflect an effort to balance national growth pressures with local risk management and ecological preservation.

Laem Chabang's example shows that even in systems with structural limitations, innovation may be facilitated by city-level leadership, academic cooperation, and a practical policy window. Even if there are still obstacles to overcome, particularly in terms of scaling initiatives and securing regulatory backing, this case shows that local governments have the ability to play a significant role in Thailand's climate governance environment.

6.5 Toward Integrated and Adaptive Governance: Lessons and Recommendations

Thailand's conflicting aims of environmental stability, economic expansion, and climate resilience are put to the test in real time in the Eastern Economic Corridor (EEC), which is more than just a zone of spatial development. This chapter's analysis of Chonburi, Rayong, and Chachoengsao has brought to light the disparity in institutional capacity distribution, the vulnerability of decentralized planning systems, and the disjointed architecture being used to implement climate governance.

It is evident from these case studies that infrastructure alone is insufficient to create climate resilience. Institutions that can manage complexity, share authority, and co-produce knowledge across sectors and scales are necessary to foster it.

Although effective in facilitating investment, the EEC Act's predominance of centralized decision-making has all too frequently marginalized local planning procedures, undermined ecological protections, and made climate adaption an afterthought. However, there are also indications of institutional innovation and resistance within this constrictive setting, such as communities organizing for ecological justice, universities incorporating research into policy, and local governments debating zoning reforms.

The conflicts between local ecological identity and national directives are best shown by Chachoengsao's opposition to top-down industrial zoning. An emerging form of place-based resilience rooted in local values and lived knowledge is demonstrated by the province's initiatives to mobilize civil society, support eco-cultural planning, and carry out independent environmental evaluations. On the other hand, Rayong's secondary cities demonstrate how inadequate planning capabilities, a lack of technical expertise, and tight financial constraints exacerbate climate risks. However, they also demonstrate how small advancements in green space design and school-led awareness initiatives can start to move governance from reactive to proactive.

Chonburi's Laem Chabang Municipality is notable as a leader in adaptation-oriented government (Pruksanubal, 2016). Its proactive use of adaptive zoning, green infrastructure, and early warning systems shows how local leadership, with the help of flexible planning windows and university partnerships, can start to balance climate-sensitive urban design with economic development. Laem Chabang shows what is possible when local governments are given the authority, resources, and willingness to take the lead, despite certain limits.

Together, these cases point toward a set of guiding principles for advancing integrated and adaptive climate governance in Thailand's coastal regions (see Fig. 6.1):

1. **Strengthen Vertical Coordination**

   Local reality and national visions must coincide for climate resilience to be effective. This entails amending the EEC Act and associated laws to mandate that the objectives of the National Adaptation Plan (NAP) be incorporated into investment and spatial planning. To allow local governments to contribute contextual information and priorities to national decisions, vertical communication mechanisms like provincial climate councils and intergovernmental planning forums should be established.
2. **Build Subnational Capacity**

   Mandates are not enough for local governments; they also require tools, resources, and training. Programs for building capacity must be adapted to actual municipal demands, such as cross-sectoral budgeting, data management, participatory scenario-building, and spatial planning in the face of climate unpredictability. Promoting cooperation between regional universities and municipalities can foster policy experimentation and create long-lasting technical support networks.
3. **Institutionalise Non-State Actor Participation**

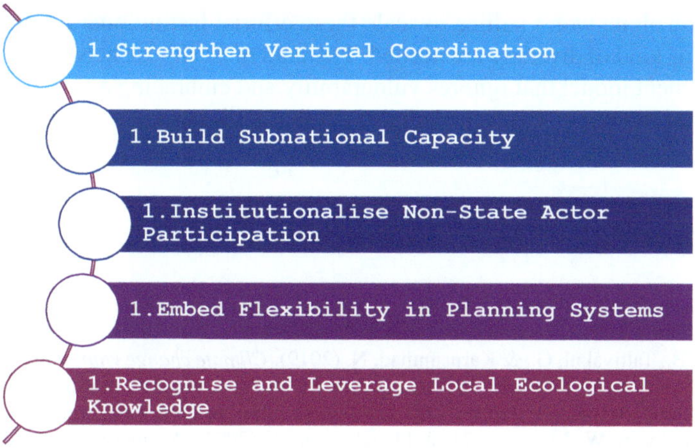

Fig. 6.1 Summary of a set of guiding principles for advancing integrated and adaptive climate governance in Thailand's coastal regions

Universities, non-governmental organizations, and private sector players are not ancillary; rather, they are vital repositories of information and catalysts for innovation. Governance tools including inclusive monitoring committees, public-academic policy labs, and adaptation task teams must be used to formalize their role. This encourages responsibility and grounds adaptation in a variety of viewpoints and well-founded knowledge.

4. **Embed Flexibility in Planning Systems**
   Dynamic and unpredictable climate futures are not well suited for rigid master plans. Land use plans must be revised on a regular basis, adaptive zoning must be permitted, and design requirements must be flexible enough to take into account new dangers. Through rewards, information exchange, and legal recognition, pilot programs ought to be integrated into larger regulatory frameworks.
5. **Recognise and Leverage Local Ecological Knowledge**
   Important insights about resilience can be gained from traditional water management methods, Indigenous ecological practices, and community-based observation. Planning procedures must give communities real opportunities to influence choices, refute prevailing narratives, and collaborate on adaptation plans that suit local environmental circumstances and values.

Finally, a paradigm shift in governance is necessary for Thailand's journey toward climate-resilient development, especially in key areas like the EEC: from blueprint planning to flexible systems thinking; from economic optimization to socio-ecological regeneration; and from hierarchical implementation to collaborative adaptation. Rethinking institutions as learning ecosystems that can handle uncertainty with humility, agility, and care is just as difficult as changing policy.

Although the Chonburi, Rayong, and Chachoengsao examples may not give a single paradigm, they do offer contextualized understandings and useful precedents.

Replication alone won't suffice to scale these efforts; institutional frameworks that support this kind of diversity are needed. Thailand must choose between entrenching a development model that ignores vulnerability and embracing a future governance model that integrates, adapts, and shares climate resilience.

References

Bhrammanachote, W. (2019). The review of Thailand's Eastern Economic Corridor: Potential and opportunity. *Local Administration Journal, 12*(1), 73–86.

Chaiyarak, B., Tattiyakul, G., & Karnsunthad, N. (2019). *Climate change vulnerability assessment Bang Pakong River Wetland, Thailand*. IUCN. X + 73pp. https://iucn.org/sites/default/files/2022-08/climate_change_vulnerability_assessment_bang_pakong_river_wetland_thailand.pdf

Charoentrakulpeeti, W., Mahawan, N., & Pharcharuen, W. (2021). An approach to enhancing local assets for green area protection—A case study on Song Khlong, Chachoengsao Thailand. *Turkish Journal of Computer and Mathematics Education, 12*(8), 2576–2585.

Cheshmehzangi, A. (2021). Public rights and maintaining the city's social and public services. In *Urban health, sustainability, and peace in the day the world stopped* (pp. 131–137). Springer Singapore.

Cheshmehzangi, A. (2025). Innovation and environmental sustainability: ICT and beach management. In *Urban innovation and sustainability: ICT-based interventions in Hainan, China* (pp. 59–82). Springer Nature Singapore.

Cheshmehzangi, A., & Dawodu, A. (2018). Stakeholder methods to climate change and sustainable development. In *Sustainable urban development in the age of climate change: People: The cure or curse* (pp. 23–68). Palgrave Macmillan.

IEAT. (n.d.). *The Industrial Estate Authority of Thailand (I-EA-T)*. https://www.ieat.go.th/en/roles-responsibilities

Jaiprasert, Y. U., & Maluleem, J. (2023). *The role of Thailand EEC to achieve ASEAN connectivity vision, aim to increasing competitiveness of Thailand economic uplifting from middle income trap* (Doctoral dissertation, Thammasat University). University Repository.

Jongtong, W. (2021). *Evaluating the land administration organization: A case of Agricultural Land Reform Office in Thailand* (Master's thesis). University of Twente.

Kamnuansilpa, P. (2014). Images of Tambon Administration Organizations in Sating Phra District of Songkhla Province, Thailand. *Journal of African & Asian Local Government Studies, 3*(3).

Kiguchi, M., Takata, K., Hanasaki, N., Archevarahuprok, B., Champathong, A., Ikoma, E., Jaikaeo, C., Kaewrueng, S., Kanae, S., Kazama, S., & Kuraji, K. (2021). A review of climate-change impact and adaptation studies for the water sector in Thailand. *Environmental Research Letters, 16*(2), Article 023004.

Kim, H., & Aroonpipat, S. (2016). *AEC preparation under recentralized Thai government: Case study of Chonburi and Rayong* (Doctoral dissertation, Thammasat University).

Kohpaiboon, A. (2020). *Industry 4.0 policies in Thailand*. https://www.iseas.edu.sg/wp-content/uploads/pdfs/ISEAS_EWP_2020-2_Archanun.pdf

Meeros, W., & Wichakul, S. (2022, October). Thailand Nationally Determined Contributions (NDCs): Analysis of climate policy for enhancing climate resilience in infrastructure construction sector. In *2022 International Conference and Utility Exhibition on Energy, Environment and Climate Change (ICUE)* (pp. 1–8). IEEE.

Molle, F., & Srijantr, T. (Eds.). (2003). *Thailand's rice bowl: Perspectives on agricultural and social change in Chao Phraya Delta*. Studies in Contemporary Thailand No. 12, White Lotus Publisher. https://horizon.documentation.ird.fr/exl-doc/pleins_textes/divers15-08/010033854.pdf

Ngernthuam, R., Soonthorn, S., Sutthisai, W., Saiyaros, R., & Boonruan, S. (2024). Evaluating key determinants of balanced development in Thailand's Eastern Economic Corridor of Rayong Province. *Pakistan Journal of Life & Social Sciences, 22*(2), 822–840.

Niyomsilp, E., Worapongpat, N., & Bunchapattanasakda, C. (2020). Thailand's Eastern Economic Corridor (EEC): According to Thailand 4.0 economic policy. *Journal of Legal Entity Management and Local Innovation, 6*(2), 219–227.

Nuamcharoen, S., & Sattakorn, W. (2021). Good governance of the provincial administrative organization in Thailand: A case study in Phrae Province, Thailand. *Psychology and Education, 58*(5), 3135–3141.

Pruksanubal, B. (2016). Land use transformation process in Chachoengsao Province, Thailand. *Procedia-Social and Behavioral Sciences, 222*, 772–781.

Secchi, C., & Belladonna, A. (2020). *Infrastructure in a changing world: Trends and challenges.* https://www.torrossa.com/it/resources/an/4698879

Srisomyong, N. (2010). *Agritourism, rural development and related policy initiatives in Thailand.* Sheffield Hallam University.

Suwanmala, C., & Weist, D. (2009). Thailand's decentralization: Progress and prospects. In *Decentralization policies in Asian development* (pp. 193–224). World Scientific.

Teerawattana, R., & Yang, Y. C. (2019). Environmental performance indicators for green port policy evaluation: Case study of Laem Chabang port. *The Asian Journal of Shipping and Logistics, 35*(1), 63–69.

Teo, H. C., Lechner, A. M., Walton, G. W., Chan, F. K. S., Cheshmehzangi, A., Tan-Mullins, M., Chan, H. K., Sternberg, T., & Campos-Arceiz, A. (2019). Environmental impacts of infrastructure development under the belt and road initiative. *Environments, 6*(6), 72.

Tongdethsri, T. (2021). *Discourses of climate change adaptation policy in Thailand: Implications for sustainable development and sufficiency economy* (Doctoral dissertation, The University of Waikato).

Tontisirin, N., & Anantsuksomsri, S. (2021). Economic development policies and land use changes in Thailand: From the eastern seaboard to the eastern economic corridor. *Sustainability, 13*(11), 6153.

Tontisirin, N., Phoomikiattisak, D., & Anantsuksomsri, S. (2017). Land use change in the eastern economic corridor of Thailand: An application of cellular automata-Makov Model. In *The 54th Annual Meeting of the Japan Section of the RSAI* (pp. 1–7).

Wongsa, S. A., Sueathung, S. U., & Tebakari, T. A. (2020, September). Climate change and adaptive water management in Bangpakong River, Thailand. In *Proceedings of the 22nd IAHR APD Congress (Sapporo, 2020), Sapporo, Japan* (pp. 14–17).

Yamsrual, S., & Potipituk, C. (2021). Furthering renewable energy for climate change mitigation: Case study on the implementation in Eco-Industrial Estates in Thailand. In *Renewable energy for mitigating climate change* (pp. 1–12). CRC Press.

Yu, Z., Deng, X., Cheshmehzangi, A., & Gao, Y. (2025). Assessing future land use demands in response to land degradation risk and Socio-Economic impacts. *Ecological Indicators, 175*, Article 113529.

Chapter 7
Community-Based Adaptation and Participatory Planning in Southern Thailand: Lessons from Ban Nai Nang Village, Krabi Province

Entry to the Chapter: Sharing My Experience of the Region
During my journeys into the Muslim communities of Krabi Province, I encountered not only a different rhythm of daily life, but also a deeper awareness of what it means to live with, and adapt to, environmental uncertainty. In villages such as Ban Nai Nang, Koh Klang Island, Baan Nateen, and the iconic Koh Panyee, I witnessed first-hand how vulnerability and resilience are not opposites, but intertwined forces shaping local governance and social life. These communities have cultivated a unique form of grounded adaptation, rooted in lived experience, traditional knowledge, and a collective ethic of stewardship. The land here seems as tired as my homeland's and hold many desires beyond just the Avatar movie scenes, showcasing a sort of environmental connectivity to the region.

In all my visits, Ban Nai Nang stood out as a quiet yet powerful example of community-based adaptation in action. Tucked between limestone cliffs and mangrove forests, the village demonstrates how participatory planning can be embedded into daily practices; be it through cooperative land management, sustainable farming methods, or the rehabilitation of wetlands. The village's use of agroforestry, organic rice cultivation, and local-led ecotourism illustrates an integrated approach to both livelihood resilience and ecological care. These are not abstract concepts; they are tangible actions sustained by a deep sense of place and cultural continuity.

As I spent time in Ao Nang and Ao Nam Mao, I also observed the stark disparities between tourism-driven wealth and peripheral vulnerability. This is somewhat different to what you could experience in the south-eastern coasts of Thailand. While resort economies may mask the exposure of these coastal zones, risk looms large, especially in the face of tsunamis, seasonal flooding, and coastal erosion (see Fig. 7.1). Even among varying social classes, environmental precarity remains a common thread and is visible in places you explore on the signs displayed, entrenched within the essence of recent history of environmental and natural disasters. What differentiates communities like Ban Nai Nang is not their insulation from risk, but their

A. Cheshmehzangi, *Thailand's Tropical Thresholds*, Urban Sustainability,
https://doi.org/10.1007/978-981-95-4975-7_7

Fig. 7.1 Signage for protection routes and safe zones against tsunami is very visible in and around Krabi's communities and neighbourhoods

collective ability to organize, plan, and respond as a social body. I felt there is a stronger sense of community here than other places in the region.

The broader demographic composition of the region, approximately 60% Buddhist and 40% Muslim, adds another layer to the complexity of adaptation in Krabi. Within this cultural mixture, Muslim communities have preserved distinctive governance practices that prioritize consensus, religious teachings, and customary land tenure. Their mosques serve not only as places of worship, but as hubs of decision-making and mutual aid. I entered a few of these local mosques and ended up witnessing at least two of these decision-making discussions. Here, participatory planning is not a policy import; it is a way of life, evolving over generations. I very much appreciated this strong sense of ownership, analytical mindset, and collective and collaborative governance.

What I found most compelling was the harmony between ecological knowledge and social cohesion. In these communities, the mangrove is not just a natural buffer; it is kin. The rice fields are not merely agricultural assets; they are ancestral landscapes. Adaptation, then, is not a reactive mechanism; it is anticipatory, embedded in the values of trust, shared responsibility, and intergenerational care.

The lessons from Ban Nai Nang and its neighbouring villages and communities highlight a powerful truth: resilience does not arrive from outside; it is grown from within. In the context of southern Thailand, particularly among its Muslim communities, community-based adaptation and participatory planning offer a grounded, dignified, and hopeful path forward amid climate disruption and development pressures. These villages remind us that adaptation is not only technical, but also cultural, emotional, and deeply collective.

7.1 Situating Community-Based Adaptation in Southern Thailand

The conceptual foundation of community-based adaptation (CBA) in the coastal environments of southern Thailand is introduced in this first section. It talks on how crucial local agency, cultural identity, and the connections between ecosystems and livelihoods are in forming adaptive strategies (Sarwono et al., 2022). It also considers the benefits and difficulties of participatory planning in the face of socioeconomic marginalization and ecological uncertainty.

With its foundation in the knowledge, habits, and agency of those most directly affected by environmental change, community-based adaptation (CBA) has become a potent alternative to top-down climate governance (Phuong et al., 2022). This strategy is a lived necessity rather than merely a strategic decision in southern Thailand's vulnerable coastal areas, where natural systems are intricately linked to livelihoods, landscapes, and identities. The daily reality of rural and peri-urban communities, coupled with rising sea levels, changing monsoon patterns, saltwater intrusion, and coastal erosion, necessitate responses that are socially inclusive, culturally rooted, and customized.

The example of Ban Nai Nang, a small Muslim-majority village in Krabi Province's Ao Luek District, shows how community-based adaptation can be used to not only address climate threats but also to rebuild ecosystems, change governance dynamics, and develop long-term adaptive capacity (Attagara, 1967). Ban Nai Nang is surrounded by karst mountains and mangrove forests, and it is situated close to delicate estuary ecosystems. As a result, it is vulnerable to tidal floods, dwindling fisheries, habitat degradation, and the socioeconomic stresses of seasonal livelihood precarity and youth outmigration. However, the community has led the way in locally driven adaptation techniques rooted in intergenerational communication, trust, and participatory planning processes—exactly in this unstable environment.

Southern Thailand's combination of cultural variety and environmental vulnerability creates a special environment for the formation of CBA (Friend et al., 2016; Kaenchan & Gheewala, 2017; Willroth et al., 2012). The Andaman and Gulf coasts, which have vast lowlands, river deltas, and agriculture that depends on the monsoon, are particularly vulnerable to climate-related risks (Becker et al., 2024). However, the area is also home to a variety of cultures, each with its own governance traditions and environmental interactions, such as Buddhist, Muslim, and Indigenous inhabitants. State-led frameworks for disaster and climate planning have frequently found it difficult to effectively address this variety, instead turning to infrastructure- or technocratic-centric strategies that ignore local perspectives.

A different strategy that emphasizes relational knowledge, group action, and flexible social structures is provided by CBA. The logic of community-based adaptation in Thailand and other countries does not deny formal government or outside research; rather, it aims to reframe local people as co-creators of resilience methods rather than as passive recipients or targets of implementation. This calls for a change from one-time involvement to ongoing cooperation—from extractive consultation to true co-design.

Deep ecological intimacy, a sensitivity to place that transcends material, spiritual, and historical dimensions, is what sets CBA apart in southern Thailand. For example, adaptability is ingrained in religious teachings, seasonal rituals, and everyday livelihood practices in Ban Nai Nang rather than existing as a stand-alone activity. Unwritten rules of harvesting, spatial memory, and respect for the natural order sustain the intergenerational and moral relationship between mangrove ecosystems and fisheries, which goes beyond just economic considerations. In this case, adaptation entails combining new technologies like digital monitoring and biodiversity mapping with more established methods like rotational fishing, selective crab harvesting, and tidal planting cycles.

Crucially, the Ban Nai Nang case also calls into question the widely held belief that community-led adaptation develops naturally or consistently in many circumstances. Instead, it shows that intentional investment in relationships, institutions, and common learning areas builds community-based resilience rather than granting it. In the last ten years, the community has established peer-to-peer learning exchanges with other coastal communities, hosted seminars with academics and planners, and fostered relationships with NGOs like RECOFTC and the Mangrove Action Project (Colchester, 2002; Erftemeijer & Bualuang, 2002; Janmaimool, 2016). Local knowledge has not been replaced by these exchanges; rather, they have strengthened the ability of the community to articulate itself (Thompson, 2018), allowing villagers to convert their experience into planning language that funders and government organizations can understand.

In a national planning setting where centralized adaptation frameworks continue to be prevalent, this is especially crucial. Although resilience is emphasized in Thailand's National Adaptation Plan (NAP) and Climate Change Master Plan, there is frequently insufficient room for truly bottom-up planning in its operationalization (Tongdethsri, 2021). Although subnational authorities are urged to include local risk data, there are still insufficient systems in place to do so, especially for small settlements located outside of urban administrative borders. The history of Ban Nai Nang indicates that the voids left by official state organizations can be filled by parallel systems of governance, such as civil society coalitions, religious institutions, and community networks.

The story also emphasizes how crucial procedure is to adaptation rather than outcome. The community has been involved in continuous processes of risk reflection, participatory mapping, seasonal scenario planning, and consensus-based action instead of delivering a single masterplan or infrastructure fix. By encouraging collaborative ownership and adaptive learning, these iterative techniques make sure that the community's reaction changes as the social and environmental realities do. As a result, resilience is woven into the village's social fabric as a dynamic, relational activity rather than a list of tasks.

The relationship between adaptation, social justice, and cultural identity is further highlighted by the southern Thai environment. The allocation of adaptive knowledge and responsibility in Ban Nai Nang is influenced by intergenerational hierarchies, gender roles, and religion beliefs. For instance, women have become important leaders in climate-resilient livelihoods, such as networks for herbal medicine and the production of mangrove honey. In the meanwhile, young people are leading the way in digital storytelling and environmental monitoring. They are active areas of change, where adaptation serves as a means of redistributing voice and altering community responsibilities in ways that promote social cohesiveness. These roles are neither static nor symbolic.

As will be demonstrated in this chapter, Ban Nai Nang's experience offers a wider range of lessons for participatory adaptation planning, including the importance of establishing trust prior to technical intervention, the effectiveness of dialogue-based planning tools like risk calendars and mapping walks, the value of resilience hubs as continuous platforms for coordination, and the requirement for institutional intermediaries who can translate between bureaucratic languages and community worldlines.

In conclusion, community-based adaptation in southern Thailand involves a complete reorganization of climate governance, not just a type of local action. It pushes politicians, scholars, and planners to embrace the lived complexity of coastal communities through co-creative engagement rather than extractive involvement. Ban Nai Nang is a solid, convincing example of how adaptation may be inclusive, culturally relevant, and genuinely transformative—when the process is anchored in people, location, and partnership—as the next sections will go more into.

7.1.1 *Context and Method: Ethnographic Fieldwork in Ban Nai Nang*

Technical examination of risks and interventions alone is insufficient to comprehend the lived experience of adaptation; communities must be deeply and firmly engaged as knowledge bearers, storytellers, and change agents. The insights offered in this chapter regarding Ban Nai Nang are based on extensive ethnographic fieldwork and participative approaches carried out during several visits between 2020 and 2024. The village's geographic and sociocultural background is described in this subsection, along with the research methodology, instruments, and moral standards that were employed to produce joint knowledge.

The little town of Ban Nai Nang is located in Krabi Province's Ao Luek District, tucked away between vast mangrove forests and karst limestone cliffs on the Andaman coast (Ruohomaki, 1997). About 300 mostly Muslim households call it home, and their main sources of income include ecotourism, small-scale fishing, crab harvesting, and rubber tapping. With tidal inlets, peat soils, and estuarine systems that sustain a wide range of marine and woodland species, the terrain is both delicate and abundant in biodiversity. The ecological cycles that support Ban Nai Nang's livelihoods have been steadily upset by changing seasonal patterns, increased saltwater intrusion, and coastal erosion, as has happened in many southern coastal settlements.

Culturally, the community is organized on close-knit kinship networks, a strong heritage of group activity, and religious leadership by the local Imam. Religious teachings and environmental ethics are closely related, and decisions about the use of natural resources are frequently decided by consensus in village meetings. Because it fosters continuity, trust, and a sense of shared responsibility for the local environment, this cultural fabric has been shown to be a crucial enabling condition for community-based adaptation.

Participatory ethnographic concepts served as the foundation for the research technique in Ban Nai Nang, which put the community at the center of knowledge creation. In order to record variance in livelihood practices and environmental conditions, fieldwork was conducted in four primary stages that corresponded with seasonal cycles. In addition to observing adaptation, the goal was to jointly establish an environment where discussion, introspection, and planning may arise via collaborative research.

The following research tools were used:

1. **Participatory Mapping and Risk Walks**

   During mapping exercises, community members worked together to locate holy biological locations, crab breeding grounds, mangrove degradation areas, and historical flooding zones. These maps were digitalized based on community input after being jointly created on large-format paper. In order to visually detect indicators of environmental change, such as damaged paths, plants stressed by salt, or abandoned fishing locations, elders and young people participated in "risk walks" through the hamlet and its environs.
2. **Seasonal Calendar Workshops**

Creating community seasonal calendars that track rainfall, tides, harvest cycles, fish migration, disease trends, and religious holidays was the main goal of these assisted workshops. These calendars served as a foundation for talking about changes throughout time and new uncertainties, as well as helping to highlight the temporal aspects of risk and adaptability.

3. **Household Narratives and Oral Histories**

   Twenty homes were interviewed in-depth, including instructors, teenage students, elderly fishermen, rubber tappers, and female heads of household. The goal of each discussion was to document the micro-histories of adaptation, including how families dealt with previous droughts or floods, how they decided to alter their means of subsistence, and what principles influenced their reactions to environmental stress. Deeper cultural layers of knowledge, such as Islamic beliefs on environmental preservation and the ethical importance of mangrove forests, were also disclosed by oral histories.

4. **Youth-Led Photovoice and Environmental Diaries**

   Twelve secondary school students participated in a photovoice project designed to document the viewpoints of the next generation. Every participant documented family adaptation techniques, local environmental changes, and future resilience visions using a smartphone. Later, these pictures were selected for a local show, which sparked discussions between generations. Additionally, young participants kept brief environmental journals in which they recorded their daily observations regarding resource use, animal behaviour, and weather trends.

5. **Resilience Hub Dialogues**

   To address important issues including water management, livelihood transitions, and mangrove restoration, the research team co-organized a number of community "resilience hub" talks, which are open village gatherings. These gatherings served as forums for group problem-solving, brainstorming, and setting priorities for neighbourhood projects. Additionally, they functioned as feedback loops wherein initial findings were disseminated to the community and either confirmed or refuted.

Gender dynamics, power interactions, and voice distribution received special consideration during the course of the study. For every interview and group activity, informed consent was acquired. To guarantee comfort and transparency, sessions with women were conducted in culturally suitable environments, frequently in homes or gardens, and included female co-researchers. To maintain confidence and match project objectives with community values, religious leaders were contacted at every stage of the research design process.

The research's co-productive spirit persisted throughout the output phase as well. The community received the findings in a variety of media, including translated summary reports, printed maps, and a brief participatory film documentary that was co-edited with young people. These products served as more than just means for distribution; they were artifacts for continuous introspection and planning, and they are currently utilized in local policy advocacy initiatives and village school activities.

Relational insight—a knowledge of how adaptation is experienced, narrated, and co-enacted across generations, genders, and roles—was what came out of this study method in addition to data. It emphasized how local governance processes, trust-building, and community rituals play a subtle yet significant role in defining what resilience actually looks like in real life. It also demonstrated the vulnerability of these systems, showing how, in the absence of consistent support, external interventions, climate shocks, or changes in leadership can destabilize local capability.

Crucially, the methodology was created to promote endogenous planning processes, raise community awareness, and elevate marginalized voices—not to extract. Instead of being viewed as an abstract "case," Ban Nai Nang was viewed as a collaborator in the common endeavour of negotiating uncertainty, with the ethnographic process turning into a form of adaptive learning in and of itself.

These concepts are integrated into particular practice areas—participatory planning, ecological knowledge, inclusivity, and innovation—in the sections that follow. However, their depth comes from this methodological and contextual foundation. Ban Nai Nang serves as a reminder that community-based adaptation is a collaborative culture that is ingrained in the environment and maintained via care, rather than merely a set of tools.

7.2 Participatory Planning in Practice: Risk Dialogues and Resilience Hubs

At its core, community-based adaptation (CBA) is a process of rethinking how decisions are made, whose knowledge matters, and what futures are envisioned as viable. Participatory planning is not just a means of interaction. Participatory planning is no longer limited to project cycles or workshops in Ban Nai Nang; rather, it has developed into a dynamic practice that is ingrained in day-to-day existence and influenced by shared responsibility, local leadership, and collective memory. This section examines risk conversations and resilience hubs, two key mechanisms that have surfaced in the village's adaptation process. Based on trust and iterative learning, both provide forums for introspection, collaboration, and group action.

7.2.1 Risk Dialogues: Mapping Experience, Memory, and Change

A series of risk dialogues—community-led discussions intended to bring to light local understandings of environmental change, past experiences of disruption, and perceptions of impending threats—formed the foundation of Ban Nai Nang's participatory planning initiatives. These discussions, which began with assistance from a nearby non-governmental organization, swiftly developed into independently organized gathering places for religious leaders, young people, and seniors to together consider the past and look to the future.

These talks were purposefully intended to be conversational and inclusive rather than technical or bureaucratic, and they were held in casual locations such under stilted homes, in prayer halls following Friday prayers, or during nighttime garden parties. Open-ended questions at the start of the procedure included: How recently did the floods arrive early? In the past ten years, what behavioural changes have you observed in crabs? Which trees once lined the creek? By asking these questions, the village created a kind of collective memory mapping by reconstructing a shared timeline of ecological and climatic changes.

Visual aids including seasonal risk calendars and hand-drawn village maps that indicated crab bank sites, flood-prone walkways, sacred ecological places, and zones of saltwater intrusion were subsequently created from this mapping. In addition to being helpful internally, these tools were also employed in discussions with scholars, local officials, and the Tambon Administrative Organization (TAO) (Meesook et al., 2020). In doing so, the community reversed the usual hierarchy of expert-driven risk evaluations and affirmed its experiential knowledge as a legitimate and essential basis for planning.

Crucially, the risk discussions also made room for intergenerational compromise. While younger people shared observations from school lessons, modern tools, and evolving goals, elders contributed long-term memory and traditional wisdom. The frequent pitfalls of nostalgia-based planning and anchored adaptation in the lived reality of all community members were avoided thanks to this convergence of viewpoints.

7.2.2 *Resilience Hubs: Anchoring Participation in Place*

The community realized the need for a more organized but adaptable setting where continuous planning and adaptation conversations could occur as risk dialogues gained traction. This resulted in the establishment of what the locals started referring to as the "resilience hub", which is not a physical structure per se but rather a series of events and activities hosted in a variety of public areas, such as the school, mosque courtyard, and mangrove learning center.

The resilience center was intended to be a multipurpose platform that served as a social anchor, a planning forum, and a workshop. Its primary responsibility was to coordinate adaptation efforts in several subject areas, including environmental education, livelihoods, water management, disaster response, and youth participation. Additionally, it provided a tool to monitor the advancement of community objectives like reforesting degraded mangrove sections, enhancing drainage along flood-prone paths, or growing crab banks.

The resilience hub's revolving facilitation approach was one of its unique features. The community implemented a shared leadership model, in which several groups alternated in leading sessions, as an alternative to depending on outside facilitators or a set committee. One month, the women's cooperative would hold a conversation about the resilience of herbal gardens; the next, the youth group might show drone footage of the health of mangroves; and the one after that, religious leaders might offer their thoughts on faith and stewardship in the face of climate change. This strategy made guaranteed that the adaption process was locally owned and diverse.

Additionally, the resilience hub developed as a crucial conduit for outside involvement. In addition to observing, provincial environmental officers, non-governmental organizations, university researchers, and donor representatives were invited to take part on an equal basis as participants, learners, and listeners. A community policy that prioritized community-defined needs, transparency of intentions, and respect for local values governed these interactions.

The resilience hub developed become a place for experimentation and creativity in addition to planning. Here, pilot projects were started, including experimenting with organic composting techniques for saline soils, employing bamboo fencing to fortify the shoreline, and testing solar-powered water pumps during dry seasons. The hub offered a secure environment for learning from mistakes and making adjustments, even when some projects were unsuccessful or needed to be rethought. Through these cycles of action and reflection, the community developed what they called *khamrop*, or 'resilience' in the local dialect: the ability to regenerate, reorganize, and evolve in addition to being resilient against shocks.

The theory of Ban Nai Nang's participatory planning approach—that planning must be relational, iterative, and imbued with cultural meaning—distinguishes it from other approaches. Resilience hubs and risk dialogues are social technologies that enable a community to choose priorities, act coherently, and make sense of change together. They are not objectives in and of themselves. They reiterate that relationships and process are just as important to adaptation as tactics and results.

Ban Nai Nang's experience provides an alternative model as adaptation becomes more and more important to Thailand's regional and national climate policy. This approach opposes extractive involvement and supports locally rooted, dialogic, and collaborative planning procedures. It demonstrates that the core of resilience is not the complexity of tools but rather the strength of internal planning, the openness of communication, and the depth of trust.

7.3 Ecosystem Knowledge and Intergenerational Adaptation Strategies

Community-based adaptation in Ban Nai Nang is based on the accumulated ecosystem knowledge that has been moulded by generations of close interaction with coastal and estuarine settings, not on outside knowledge or imported models.

This knowledge is a dynamic, living system of knowing that directs how the community predicts, responds to, and recovers from environmental change. It is based on social practice, religious ethics, and everyday observation.

A key component of this system is the mangrove forest. Villagers see the forest as a "protector of life"—a sacred area that supports biodiversity, stabilizes shorelines, and offers food, medicine, and spiritual significance—in addition to being a natural barrier against coastal storms. Elders cite minor changes in tidal rhythms and water salinity as early warning signs of disturbance, and they recall the "sound of the crabs" at moonlit tides as an indication of ecological health. The village's ability to adapt depends on this fine-grained ecological literacy, which is transmitted orally and through embodied actions.

For example, fishing families preserve marine regeneration by avoiding breeding periods and using tidal calendars based on lunar cycles to indicate safe harvesting times. Local conversations about diminishing catch quantities and the loss of keystone species led to the decision to create community-run crab banks, which are protected enclosures for females that lay eggs. The village was able to put into practice an ecosystem-based approach that satisfies the needs of livelihood and conservation by utilizing traditional resource management techniques.

This knowledge system's connection with local religious leadership and Islamic beliefs is another important feature. During Friday sermons, the Imam of Ban Nai Nang frequently ties religious ethics to environmental care by using passages from the Quran regarding the sanctity of the soil and the moral obligation of stewardship (khalifa). Rituals like planting trees or cleaning up mangroves are frequently incorporated into religious celebrations, such communal get-togethers after Ramadan, combining social cohesiveness, ecological awareness, and faith. The community's adaptation techniques are strengthened by this religious framing of environmental stewardship, which grounds them in a moral and spiritual logic that fortifies accountability and participation.

However, this body of knowledge is neither static nor sentimental. It develops by experimentation, intergenerational communication, and the intentional involvement of young people in environmental education. Ban Nai Nang has made investments in youth-led educational programs that integrate traditional and modern modes of knowledge. Older fishermen, for instance, tell stories in the village school about how shrimp behave during periods of intense rain or the characteristics of soils that are ideal for growing nipa palms. These tales are not merely anecdotal; rather, they serve as predictive models, full of valuable insights that assist newer generations in understanding change and modifying their actions accordingly.

With the help of collaborations with nearby universities and non-governmental organizations, students also participate in drone-based mangrove canopy research, digital mapping, and biodiversity surveys. In one project, secondary school students took pictures of their families' seasonal pursuits, such as fishing, honey harvesting, and rubber tapping, and then captioned the images to show how each activity reacts to shifting climate patterns. These diaries developed into instruments for group introspection, igniting debates within the village resilience hub and igniting family interactions. Along with contributing to a digital community archive, the young people save environmental data, maps, and oral histories for later use.

A multidirectional learning environment where each generation teaches and learns has been developed by combining the lived wisdom of elders with innovation led by young people. Moreover, it makes adaptive experimentation easier. For instance, using their elders' understanding of groundwater movement and plant behaviour, younger community members tested bamboo-based water retention pathways after noticing an increase in the frequency of dry periods. At resilience hub meetings, these micro-infrastructure experiments were presented, modified in response to input, and occasionally expanded to other areas of the village.

It is impossible to overestimate the importance of the family in this transmission of knowledge. Adaptive behaviours are fostered through everyday activities like cooking, gardening, gathering firewood, and trash management; they are not limited to formal schooling or public gatherings. Despite being informal and frequently gendered, these "micro-adaptations" are effective learning environments. Grandfathers educate grandsons how to predict wind direction by listening to the rustle of nipa fronds or the waterbirds' migration patterns; women teach their daughters about storm-resistant farming methods or herbal cures for heat-related ailments.

The foundation of community resilience is made up of these routine behaviours. They make it possible for adaptation to be both individual and group-based, ingrained in both the private rhythms of life and plans and policies. They also aid in preventing the loss of information that might happen amid quick changes in the socioeconomic landscape. The community's investment in documenting, disseminating, and advancing its ecological knowledge base becomes a crucial continuity strategy as more young people migrate for employment or study, or as development changes the landscape and patterns of life.

This knowledge system is not risk-free, though. Intergenerational transmission is at risk from formal education curriculum that marginalize local environmental knowledge, encroaching industrial expansion, and increasing tourism pressures. Villagers are worried that if they are not actively taught, their knowledge of crab breeding cycles and how to identify therapeutic plants may disappear. The community has started incorporating traditional knowledge modules into the programming of resilience hubs and school environmental clubs to solve this. Youth are explicitly acknowledged as "knowledge keepers" and elders as "community ecologists" (Forsyth & Walker, 2014), highlighting shared responsibility and reciprocal tasks.

In conclusion, intergenerational continuity and deeply ingrained environmental knowledge are the main sources of Ban Nai Nang's adaptability. The community views adaptation as a shared, changing process, where learning never stops and the environment serves as both a teacher and a partner, rather than categorizing knowledge into old and new, traditional and modern. This combination of digital experimentation, storytelling, ritual, and observation provides a model of resilience that is not just technically solid but also culturally and socially relevant.

The Ban Nai Nang story demonstrates that effective adaptation relies not only on data or infrastructure but also on memory, mentorship, and intergenerational learning as climate threats continue to escalate throughout southern Thailand. In this instance, knowledge resilience and the community's ability to preserve it collectively and adaptively are inextricably linked to place resilience.

7.4 Inclusion, Identity, and Innovation in Community Resilience

In Ban Nai Nang, community-based adaptation is profoundly social as well as ecological. The village's resilience planning strategy demonstrates a dedication to inclusivity as a cultural value ingrained in daily life rather than as a formality. The distribution of care and labour in the community, who drives innovation, and who participates in adaption decisions are all influenced by the intersection of gender roles, religious identity, and generational duties. When deliberately fostered and reinterpreted in adaptive processes, these social institutions have transformed from being obstacles to sources of strength.

The active participation of women in adaptation initiatives, especially in the areas of livelihood diversification and community care, is one of the most obvious aspects of inclusion in Ban Nai Nang (Dawson et al., 2023). Important climate-resilient endeavours like small-scale ecotourism projects, salt-tolerant herbal gardens, and the production of mangrove honey are led by women. These endeavours serve as resilience's soft infrastructure and are closely linked to the ecosystem's health, making them more than just sources of revenue. In order to provide adaptive healthcare without depending on outside systems, the women's herbal gardens, for instance, contain plants that reduce heat stress and cure waterborne infections.

Religious and cultural values play an enabling role here. Islamic beliefs on humility, stewardship (khalifa), and mutual aid (ta'awun) influence how responsibilities are understood, and authority is allocated in Ban Nai Nang. Despite the persistence of old gender conventions, women are nevertheless able to lead adaptation initiatives and take part in planning discussions. Rather, they establish a framework in which environmental protection, family guidance, and group responsibility are ways that leadership is demonstrated. In this situation, women's leadership is relational rather than positional; it is a type of authority based on competence, trust, and service.

Additionally, the care economy, which is frequently overlooked in mainstream planning, is intertwined with adaptation. Women oversee water collection, food security, and the care of children and the elderly, all of which are significantly impacted by climate unpredictability. Ban Nai Nang guarantees that adaptation methods are sensitive to the entire range of lived vulnerabilities, not just those identified in technical assessments, by establishing inclusive planning spaces that value these perspectives. The legitimacy, ownership, and long-term viability of community resilience initiatives are all improved by this inclusivity.

In Ban Nai Nang, youth involvement has also increased dramatically, with young people assuming roles that defy conventional wisdom. They are increasingly considered active participants in innovation, communication, and information creation rather than merely future stakeholders. Youth groups conduct digital literacy workshops for peers and elders, plan mangrove education excursions for tourists, and take part in biodiversity tracking. These responsibilities are frequently balanced with societal duties, family work, and religious education to produce hybrid identities that unite tradition and change.

Crucially, the involvement of women and youth is viewed as a strategic imperative for climate resilience rather than as a tokenistic endeavour. The social structures that mediate adaptation must change along with the speed and unpredictability of climate impacts. The experience of Ban Nai Nang demonstrates that inclusive adaptation planning involves more than just who is present; it also involves recognizing and incorporating many approaches to leadership, caring, and knowing into the process of making decisions. The village provides a nuanced example of socially embedded resilience in this way, redefining tradition to meet current and future demands while allowing for new internal leadership, voice, and creativity.

7.4.1 Youth and Digital Tools: Monitoring Change from the Ground Up

The youth-led use of digital tools in Ban Nai Nang has greatly influenced the village's adaptation process, transforming social media platforms, mapping apps, and cell phones into tools for community storytelling, environmental monitoring, and climate education. A distinctive form of bottom-up climate intelligence is reflected in this combination of digital innovation and traditional ecological knowledge, in which young people work as intermediaries between ecosystems, generations, and outside actors.

A small group of secondary school kids started experimenting with smartphone photography to record changes in the mangrove forest, including dieback regions, illegal dumping, and tidal shifts. This was the beginning of this digital involvement. After a local NGO led a youth training series on participatory mapping and environmental monitoring, this once unstructured activity took on a more official form. The young people started recording crab habitats, bird sightings, and erosion-prone routes using open-source biodiversity tools and simple GPS-enabled applications. They soon became well-known in the village and among outside researchers for their work.

One significant innovation was the creation of a community biodiversity database, which was carried out by the youth group under the supervision of elders who provided local species names and taxonomies. Youth members cross-checked species health indicators, geotagged photographs of observations, and even highlighted local uses for various plants and animals using platforms like iNaturalist and KoboToolbox (Atsumi et al., 2025; Donnelly, 2024; Wichaidit, 2024). Teachers, tour guides, and visiting scientists all used this digital ecosystem record as a teaching tool in addition to other young people.

Through a collaboration with a nearby university, students also started tracking the condition of mangroves using basic drone flights. They started tracking canopy cover, sedimentation, and indications of forest fragmentation throughout the seasons after receiving instruction in aerial photography. This information was incorporated into conversations about reforestation goals and visualized into community maps that were on display at the resilience hub. Drone photography also aided in the development of visual narratives that local officials, kids, and the elderly could all understand regarding complicated environmental change.

Participatory media and storytelling have been a significant aspect of youth-led digital engagement. Through the creation of the Voices of the Mangrove community Facebook page and short video series, the youth group chronicled seasonal cycle rituals, demonstrated adaptive farming practices, and conducted interviews with elders. In order to bridge generational gaps and broaden the audience for Ban Nai Nang's adaptation story, these digital stories—which were frequently created in both Thai and the regional Malay dialect—were distributed online and at village get-togethers.

Through these initiatives, the village's kids are now seen as active stewards of the environment rather than passive consumers of adult knowledge. Although elders still possess profound ecological knowledge, young people are now acknowledged as climate interpreters who can communicate hazards in a way that appeals to their younger peers, establish connections with tech-savvy organizations, and interact with outside networks.

These digital tools have affected actual planning decisions in addition to communication. For example, the municipality decided to upgrade drainage systems along a crucial walkway based on flood mapping done by the youth group utilizing field surveys and open-source satellite photos. The Tambon-level environmental reporting platform has also received data on biodiversity created by young people (Thitiprasert et al., 2007), which has an impact on how conservation resources are allocated.

But there are some conflicts with this digital interaction. Concerns over data privacy, deception, or an excessive dependence on technology at the expense of in-person observation were voiced by a few elderly peasants. In response, the youth group organized community validation sessions where digital discoveries were reviewed and rectified in public settings, reinforcing the idea that digital technologies should support community authority and lived experience rather than take their place.

In conclusion, the young people of Ban Nai Nang show that adaptive innovation is not limited to infrastructure or funding. It lies in the capacity to notice, connect, and tell the story of transformation from the ground up. Their use of digital technologies has improved intergenerational communication, broadened the community's adaptive knowledge base, and established the village as a center for grassroots climate intelligence. It is a striking illustration of how technology may be a potent ally in fostering inclusive, long-term resilience when it is rooted in local relationships and values.

7.5 Reflections and Recommendations: Process, Trust, and Long-Term Capacity

The tale of Ban Nai Nang village is one of gradual, methodical, and profoundly human adaptation processes rather than one of technological innovations or impressive infrastructure. Perhaps more effectively than any external model, this story illustrates that relationships, not templates, are the reason community-based adaptation (CBA) works. In vulnerable coastal settings, the validity and longevity of adaptive solutions are determined by the process rather than the final product.

Trust has been the unifying factor throughout Ban Nai Nang's history, not only inside the community but also throughout generations, knowledge systems, and locals and outside partners. This trust was developed gradually via candid communication, shared experiences, and an iterative approach to problem-solving; it was not granted. The effectiveness of participatory approaches, such as risk dialogues and resilience hubs, was due to their respect for the local rhythms and realities rather than their technical prowess. Decisions were not made in a vacuum, but rather through procedures that honoured a range of opinions and changing requirements.

Resilience is relational before it is technical, which is a crucial takeaway from this chapter. A well-balanced social ecology underpins Ban Nai Nang's adaptability, with women maintaining the relational infrastructure of care and continuity, youth experimenting with tools and formats, and seniors leading with memory and wisdom. Here, adaptation is a practice of collective attentiveness that is continuously refreshed across time, roles, and spaces rather than an externally imposed intervention.

Other communities may learn a lot from intergenerational leadership, which is demonstrated by the mentorship relationships and the linking of traditional and digital knowledge. Resilience cannot be left to the next generation or formal institutions alone. It needs to be co-owned and shared. According to Ban Nai Nang's concept, positions are co-evolved rather than prescribed, and information is shared through friendship and teamwork in addition to instruction.

Both a challenge and an opportunity are presented by this scenario for regional and national climate policy frameworks. The difficulty derives from policy systems' propensity to place a higher priority on speed, measurement, and homogeneity. In contrast, CBA necessitates institutional humility, contextual sensitivity, and patience. It must be sown, nurtured, and permitted to develop naturally within location-specific circumstances (Xie et al., 2019); it cannot be scaled like infrastructure.

Still, the opportunity is great. CBA can provide a reproducible yet adaptable route to resilience if governments and funders change their focus from delivery to facilitation and from control to enabling environments. The experience of Ban Nai Nang yields a number of recommendations. Instead of concentrating just on tangible results, planning frameworks should incorporate time, resources, and space for inclusive processes like workshops, dialogues, and local facilitation. The first step is to institutionalize support for process-oriented adaptation. These are foundational, not peripheral.

The second step is to identify and support intergenerational learning systems, which means that laws for climate adaptation should incorporate clauses for knowledge documentation, mentorship programs, youth-led monitoring, and advisory structures with elder participation. It is important to view both formal and informal education as equally important sources of resilience. Establishing safe areas for community self-governance comes in third. In areas that are developing quickly, such as the coastal regions of southern Thailand, policies must protect populations from processes of performative or exploitative involvement (Li et al., 2021). Community voice can be protected from appropriation or dilution through mechanisms including rights-based safeguards, negotiated partnerships, and community protocols.

Fourth, connecting regional planning with local initiatives should be the main goal (Cheshmehzangi & Tang, 2024). We shouldn't ignore the lessons learned from Ban Nai Nang. Without losing their distinctiveness, local innovations can be translated into regional policy insights with the aid of tools like open data platforms, digital knowledge archives, and local–provincial interaction forums. Finally, it is critical to promote adaptability as a long-term capacity rather than a short-term endeavour. Funding arrangements should prioritize community-driven, multi-year programs over short cycles in this respect. Developing adaptive ability is a continuous process that calls for long-lasting connections, steady support, and the flexibility to try new things and grow.

Finally, Ban Nai Nang provides politicians, academics, and adaptation planners with a powerful message: resilience increases when individuals are empowered to drive change on their own. It flourishes in environments when a community's moral commitments, social ties, and cultural fabric are all interwoven with adaptation (Zahnow et al., 2025). Continuity, attention, and the bravery to stick with the process even when results are questionable are more important than perfection.

The teachings of Ban Nai Nang urge us to reframe adaptation as a practice of remaining grounded, learning from one another, and creating futures that respect people and place rather than as a race against time as the climate catastrophe worsens. We discover inspiration and a model for the kind of adaptation the world sorely needs, i.e., adaptation that listens, adapts, and endures, in the quiet strength of this town in southern Thailand.

References

Attagara, K. (1967). *The folk religion of Ban Nai, a hamlet in Central Thailand*. Indiana University.

Atsumi, K., Tsujino, T., Nakaseko, M., Ogasawara, S., Tsuboi, R., Taga, H., & Nishida, T. (2025). Crowdsourced Wildlife Observations Uncover Ecotourism Assets—Sites and Species. *bioRxiv*, 2025-05.

Becker, M., Seeger, K., Paszkowski, A., Marcos, M., Papa, F., Almar, R., Bates, P., France-Lanord, C., Hossain, M. S., Khan, M. J., & Karegar, M. A. (2024). Coastal flooding in Asian megadeltas: Recent advances, persistent challenges, and call for actions amidst local and global changes. *Reviews of Geophysics, 62*(4), e2024RG000846.

Cheshmehzangi, A., & Tang, T. (2024). Shenyang: One of China's Major Industrial Hubs with Strategic Positioning and Regional Influence. In *China under construction: Shaping cities through recent urban transformation* (pp. 177–196). Springer Nature Singapore.

Colchester, M. (2002). *Learning Lessons from the RECOFTC Experience with Community Forestry Networking*. Ford Foundation and DFID (unpublished document).

Dawson, N., Bhardwaj, A., Coolsaet, B., Massarella, K., Scherl, L., Yannick, N., Oliva, M., Suich, H., & Worsdell, T. (2023). Towards more equitable and effective nature conservation led by Indigenous peoples and local communities. *Policy Matters, 23*, 13–31.

Donnelly, D. (2024). *Biodiverse Thailand: Conservation for species at risk*. Published online as an unpublished document (17 pages).

Erftemeijer, P. L. A., & Bualuang, A. (2002). Participation of local communities in mangrove forest rehabilitation in Pattani Bay, Thailand: learning from successes and failures. In *Strategies for wise use of wetlands: Best Practices in Participatory Management. Proceedings of a workshop held at the second international conference on wetlands and development (November 1998, Dakar, Senegal)* (Vol. 56, pp. 27–36). Wetlands International, IUCN. WWF Publication.

Forsyth, T., & Walker, A. (2014). Hidden alliances: Rethinking environmentality and the politics of knowledge in Thailand′ s campaign for community forestry. *Conservation and Society, 12*(4), 408–417.

Friend, R., Choosuk, C., Hutanuwatr, K., Inmuong, Y., Kittitornkool, J., Lambregts, B., Promphakping, B., Roachanakanan, T., Thiengburanathum, P., Thinphanga, P. and Siriwattanaphaiboon, S. (2016). *Urbanising Thailand: Implications for climate vulnerability assessments*. Thailand Environment Institute.

Janmaimool, P. (2016). The establishment of a community-based mangrove Forest management plan: Lessons learned from mangrove Forest conservation in the Nernkhor sub-district, Rayong Province, Thailand. *Applied Environmental Research, 38*(3), 59–76.

Kaenchan, P., & Gheewala, S. H. (2017). Budget constraint and the valuation of environmental impacts in Thailand. *The International Journal of Life Cycle Assessment, 22*(11), 1678–1691.

Li, L., Cheshmehzangi, A., Chan, F. K. S., & Ives, C. D. (2021). Mapping the research landscape of nature-based solutions in urbanism. *Sustainability, 13*(7), 3876.

Meesook, K., Routray, J., & Ahmad, M. (2020). Rural local government finance and its management in Thailand: Reflections and prospective through Tambon administrative organisations. *International Journal of Rural Management, 16*(2), 199–224.

Phuong, N. T. N., Tran Ba, D., & Do Dieu, L. (2022). Assessing the potential of community-based ecotourism toward sustainable development: A case study in Tua Chua Karst Plateau–Dien Bien–Viet Nam. *Geo Journal of Tourism and Geosites, 44*(4), 1300–1305.

Ruohomaki, O. P. (1997). *Livelihoods and environment in Southern Thai maritime villages*. University of London.

Sarwono, A. W., Kherenhapukh, W., & Rachma, A. (2022). Mangrove forest tourism development strategy: A study case of Nai Nang Village Thailand. *International Journal of Tourism, 1*(2), 7–17.

Thitiprasert, W., Ratanasatien, C., Chitrakon, S., Watanesk, O., Chotechuen, S., Forrer, V. S., Sommut, W., Somsri, S., Samitaman, P., & Changtragoon, S. (2007). *Country report on the state of plant genetic resources for food and agriculture in Thailand (1997–2004)* (pp. 18–32). Department of Agriculture.

Thompson, B. S. (2018). The political ecology of mangrove forest restoration in Thailand: Institutional arrangements and power dynamics. *Land Use Policy, 78*, 503–514.

Tongdethsri, T. (2021). *Discourses of climate change adaptation policy in Thailand: Implications for sustainable development and sufficiency economy* (Doctoral dissertation, The University of Waikato).

Wichaidit, W. (2024). History of Dengue hemorrhagic fever-like symptoms, care-seeking behaviors, and household conditions among residents of Wang Wiset District, Trang Province, Southern Thailand. *Procedia of Multidisciplinary Research, 2*(9), 6–6.

Willroth, P., Massmann, F., Wehrhahn, R., & Revilla Diez, J. (2012). Socio-economic vulnerability of coastal communities in southern Thailand: The development of adaptation strategies. *Natural Hazards and Earth System Sciences, 12*(8), 2647–2658.

Xie, L., Tan-Mullins, M., & Cheshmehzangi, A. (2019). Political ecology of Chinese smart eco-cities. In *Remaking sustainable urbanism: Space, scale and governance in the new urban era* (pp. 57–78). Springer Singapore.

Zahnow, R., Yousefnia, A. R., Hassankhani, M., & Cheshmehzangi, A. (2025). Climate change inequalities: A systematic review of disparities in access to mitigation and adaptation measures. *Environmental Science & Policy, 165*, Article 104021.

Chapter 8
Between Sea and Sprawl: Flood Risk and Climate Resilience in the Hua Hin–Cha-Am Coastal Corridor

8.1 The Corridor in Transition: Setting the Scene

The Hua Hin–Cha-Am coastal corridor, which stretches along the northern reaches of the Gulf of Thailand, is a rapidly changing landscape, with resort development, upscale real estate, and escalating infrastructure displacing the sleepy charm of small fishing towns. Over the past 20 years, this corridor, which crosses portions of the provinces of Phetchaburi and Prachuap Khiri Khan, has become a destination for new regional development initiatives as well as a recreational area for Thailand's urban elite (Husa et al., 2014). The region is experiencing a significant demographic, economic, and spatial transformation because to its improved transportation infrastructure, close proximity to Bangkok, and reputation for pristine seaside beauty.

Yet beneath the surface of this transformation lies a convergence of risks. The corridor is particularly vulnerable to the combined stresses of coastal climate vulnerability and yearly flooding, which are made more dangerous by the slow growth of low-lying urban development. Questions of long-term resilience have become more urgent as urban footprints spread both inland and out to sea, frequently onto wetlands, floodplains, and reclaimed coastal areas (Kityuttachai et al., 2013). Once sporadic, floods are becoming more common and destructive, frequently interfering with transportation, causing property damage, and affecting livelihoods. In the meantime, the region's underlying biological stressors have started to get worse due to sea level rise, saltwater intrusion, and shifting rainfall patterns.

Understanding Hua Hin-Cha-Am's background is essential to comprehending its current climate vulnerability. Cha-Am has a good intention to develop itself alongside the fame of Hua Hin, while remaining a unique part of the province. These were originally sleepy seaside communities that relied on domestic travel, small-scale farming, and artisanal fishing. Long before modern development began, Hua Hin, long regarded as Thailand's first beach resort and a royal retreat, had cultural and symbolic importance. Locals and tourists from Bangkok used Cha-Am, which was more subdued and tranquil, as a seasonal haven. But now, both cities are part of a

A. Cheshmehzangi, *Thailand's Tropical Thresholds*, Urban Sustainability,
https://doi.org/10.1007/978-981-95-4975-7_8

larger story of state-driven modernization, speculative real estate, and tourism-driven expansion (Unhasuta et al., 2021). Along with changing the coast's physical shape, this story has also altered its ecological balance and social structure.

8.1.1 Landscapes of Change: Ecology, Economy, and Expansion

The corridor's transformation is visually striking. Gated housing estates have replaced coastal wetlands. Open agricultural fields have given way to golf courses, opulent homes, and apartment complexes. In addition to increasing land values and promoting high-intensity development, infrastructure corridors such as roads, railway renovations, and potential high-speed rail lines are speeding up connectivity with Bangkok and beyond.

These alterations have significant ecological ramifications. Because of its low-lying terrain, the area is prone to surface water buildup during periods of intense monsoon precipitation. Previously controlled by mangroves, retention areas, and natural drainage channels, uncontrolled construction and land infill have made these flood paths more difficult to use or more dispersed. Unable to percolate or disperse, seasonal runoff frequently collects in metropolitan areas, overtaxing drainage systems and causing frequent flash floods.

Coastal dynamics are changing concurrently. Due in part to sediment starvation and land subsidence, the Gulf of Thailand is already seeing a relative sea-level increase that is higher than the global norm. Where once-dynamic shorelines are now hardened by concrete interventions, erosion hotspots are appearing, especially around seawall-protected sections. Artificial changes to the shoreline have changed the wave movement and sediment flow in the Khao Takiab area of Hua Hin and the beachfront in Cha-Am, resulting in beach thinning and structural concerns (Sealeum et al., 2007). These concerns have grown significantly over the years.

Economically, there are now significant geographic divisions because of the transition to tourist and second-home economies. While the seaside resorts and condominiums are populated by wealthy domestic and foreign visitors, many service workers and informal communities are located on marginal ground further inland, frequently on the most flood-prone terrain. Strong flood resilience and inclusive planning have not always resulted from the inflow of new funding. In actuality, coordinated regional solutions to common environmental threats (Thanarotrungrueang et al., 2022) are frequently hampered by the disjointed administrative borders between the provinces of Phetchaburi and Prachuap Khiri Khan.

This change is accompanied by social friction. Traditional communities have had their land taken or used for other purposes, sometimes without proper compensation or consultation, particularly smallholder farmers and fishermen. Longtime inhabitants are becoming increasingly disengaged from the decision-making processes about the places they call home as a result of the redefining of cultural landscapes due to outside pressures and speculative interests. The corridor's risk profile is further weakened by these dynamics, which include socio-spatial marginalization in the face of climate concerns in addition to environmental exposure.

Therefore, the Hua Hin–Cha-Am corridor is more than just a location of physical change; it is a threshold area where climatic unpredictability, ecological fragility, and economic ambition meet. Managing flooding and improving infrastructure are no longer the only challenges as development continues to push against environmental limits. The goal is to rethink growth without increasing vulnerability, uprooting the weak, or destroying the natural systems that formerly protected this section of the shore. In-depth discussions of these issues are provided in the parts that follow, which also examine the changing flood dynamics, difficult policy decisions, and strategies for constructing a more climate-resilient future for one of Thailand's most iconic coastal corridors.

8.2 Flood Dynamics and Climate Pressures in a Dual-City Stretch

In the Hua Hin–Cha-Am corridor, flooding is rapidly becoming into a chronic and escalating climate concern rather than a seasonal annoyance. The region's real estate markets, tourist-driven economy, and beautiful beaches are all growing, but the underlying hydrological vulnerabilities are getting worse. This dual-city coastal region, which spans the provinces of Phetchaburi and Prachuap Khiri Khan, is vulnerable to flooding due to a combination of human-caused hydrological system disruption and natural climate stresses. Assessing the region's capacity for climate resilience and the threats that unrestrained growth may increase requires an understanding of these dynamics.

The corridor's low-lying terrain and convergent water sources are the main causes of its flooding problems. Because of the shallow slope of the ground, significant seasonal rainfall from the southwest monsoon swiftly builds up in the developed environment, while inland drainage from the Tenasserim Hills flows eastward toward the Gulf of Thailand. Coastal surge occurrences, especially during high tide or tropical storm activity in the Gulf, make both pluvial (rainfall-based) and fluvial (river and stream-based) flooding events more frequent.

Seasonal overburdening of local drainage systems, particularly in quickly urbanizing areas, is one of the region's distinguishing hydrological features/characteristics. Concrete roads, paved developments, and tiled rooftops are examples of impermeable surfaces that have developed, reducing the land's natural ability to absorb and control water flow. Due to uncontrolled land reclamation and encroachment, natural flood storage areas such riparian buffers, floodplains, and wetlands have gradually disappeared or deteriorated. Localized flooding can arise from even mild rainfall due to the enhanced surface runoff effect.

8.2.1 Rainfall Extremes and Changing Climate Norms

The Hua Hin–Cha-Am corridor has not been immune to the effects of global warming on Thailand's changing climate patterns. Extreme rainfall intensification episodes, both in terms of total volume and precipitation rate, have occurred in the region in recent years. Because of increased atmospheric warming, which permits more moisture retention and abrupt release in storm systems, these high-intensity rainfall events—which can occasionally dump more than 100 mm in a few hours—are now statistically more likely to occur.

Additionally, the seasonal pattern is growing increasingly unpredictable. There is a change in the onset and retreat pattern of the conventional monsoon, which usually lasts from May to October. Flash downpours are interspersed with erratic dry intervals. For instance, Hua Hin and Cha-Am both saw heavy rainfall events outside of their typical flood windows in 2021 and 2022. This resulted in a brief urban inundation that lasted for several hours and caused disruptions to trade, tourism, and transportation.

Municipal and provincial stormwater systems are under a lot of stress due to the combination of urban development and rainfall variability. Adequate stormwater retention infrastructure and detention ponds are lacking in several parts of Hua Hin, notably Khao Takiab and portions of the inner city. The capacity of drainage systems to quickly reroute floodwaters is diminished by their frequent narrowness, partial obstruction, or poor maintenance. The scenario in Cha-Am is comparable; during periods of high rainfall, flooding frequently occurs along Phetkasem Road and in residential areas. These difficulties are exacerbated when high tides and heavy rainfall occur together, causing drainage canals to backflow and protracted flooding.

8.2.2 Coastal Surges, Sea-Level Rise, and Erosion

The Gulf of Thailand is experiencing a relative sea-level rise estimated at more than 3 mm per year—higher than the global average—due to a combination of oceanic expansion and land subsidence from excessive groundwater extraction, which increases the risk of tidal flooding and storm surges, especially in low-lying beachfront zones, hotel strips, and reclaimed lands near the shoreline (Arunothaipipat, 2011). In addition to surface flooding, the corridor is becoming more vulnerable to coastal flooding and sea encroachment.

There have been instances of saline water intrusion into agricultural areas and coastal aquifers in Cha-Am, especially during high tide or dry season groundwater depletion. This affects inland towns who depend on shallow wells or groundwater reservoirs for their long-term water security in addition to coastal agricultural and soil salinity.

Coastal erosion is also becoming a serious risk element that goes unnoticed (Cheshmehzangi et al., 2024). Although they are designed to safeguard important infrastructure or valuable real estate, seawalls and artificial embankments frequently interfere with the sand's natural littoral drift and worsen erosion in nearby areas. Sections of Hua Hin beach are a prominent example, where coastal hardening has caused sand loss, narrowing beaches and putting boardwalks and hotels closer to the water. Because protective beach buffers are being reduced, these situations increase the likelihood of coastal flooding during king tides or cyclonic disturbances.

8.2.3 Urban Form, Informality, and Exposure Inequities

Informal and peri-urban populations inside the corridor are frequently the most vulnerable to the effects of floods (Baye, 2025), even though tourist-centric places may receive the majority of infrastructure investment and flood mitigation attention. Flood exposure is often higher on marginal terrain, such as canal sides, former agricultural plots, or backfilled marshes, where many low-income settlements have established themselves. In terms of climate adaption strategies, public safety resources, and drainage infrastructure, these communities are usually underserved.

These areas are frequently the most affected by floods, with homes submerged, access roads closed, and livelihoods interrupted. Furthermore, these places' demands are usually disregarded in mainstream flood response or catastrophe risk reduction plans because they could not have official land titles or be integrated into municipal planning processes.

The governance mismatch along the corridor makes the problem worse. Coordinated flood planning and climate adaptation are hindered by the fragmented institutional landscape created by the administrative separation between Phetchaburi (Cha-Am) and Prachuap Khiri Khan (Hua Hin). Although Hua Hin Municipality has made an effort to put local flood response plans into action, such as community warning systems, drainage upgrades, and embankment restorations, these initiatives frequently end at jurisdictional lines. Due to its distinct provincial structure, Cha-Am can have delays in planning or funding, which would result in haphazard reactions and redundant work.

8.2.4 Cascading Risks and Systemic Interactions

The cascading array of dangers is what makes flooding in this region so complicated. In addition to flooding roadways, a significant rainfall storm can cause sewerage system failures, overload the infrastructure supporting the local electricity supply, impede real estate transactions, and affect tourism. The economic and reputational risks are considerable for a corridor that depends so heavily on image, mobility, and service delivery (Tanaboriboon, 2022). For instance, during floods, hotels and resorts along Hua Hin's shoreline may experience not just structural water damage but also business disruptions from closed roads or cancelled reservations. Local workers, many of whom reside farther inland in flood-prone informal settlements and depend on daily transportation to tourism jobs, are also impacted, in addition to property owners.

Furthermore, flooding brought on by climate change may interact with hazards to public health, especially in crowded regions with poor sanitation. Outbreaks of water-borne illnesses like leptospirosis or diarrheal infections, as well as vector-borne diseases like dengue, can result from stagnant floodwaters. These health hazards increase social vulnerabilities and put additional burden on public services, particularly for migratory worker, elderly, and kid populations.

8.2.5 *A Pressing Case for Integrated Resilience Planning*

This hydrological and climatic description makes it abundantly evident that the Hua Hin–Cha-Am corridor requires a coordinated flood management policy that takes climate change into account. This strategy should overcome regional divisions and balance ecological preservation with economic development. Post-flood cleanup and ad hoc drainage upgrades are no longer adequate. Future development's spatial logic, the materials chosen, zoning laws, and the regular operations of urban systems must all incorporate resilience.

Anticipatory action is necessary due to the current exposure profile, which includes climate variability, subsiding land, weaker coastal buffers, flood-prone inland developments, and overburdened drainage. This includes upgrading stormwater infrastructure, investing in nature-based flood mitigation strategies (such wetland retention zones or mangrove restoration), improving hydrological modelling capabilities, and creating participatory early warning and evacuation systems.

Above all, ensuring flood resilience in the Hua Hin–Cha-Am corridor is a social equity, planning, and governance issue as well as a technological one. Without action, the very waters that the coastal region aims to embrace could threaten the dream of a thriving, connected, and liveable area. The planning conflicts and development conundrums that are making the region's adaptive response even more difficult are discussed in the next section.

8.3 Development Dilemmas and Planning Gaps

The Hua Hin–Cha-Am route is situated at the intersection between vulnerability and ambition. The rate of change frequently surpasses the planning frameworks intended to control it as new constructions swarm the shoreline, ranging from upscale condominiums and golf resorts to mixed-use residential estates and tourism complexes. The result is a pattern of reactive, disjointed, and disjointed urban growth that increases the corridor's vulnerability to climate and water hazards. The problem is not just about what is being built, but also about how, where, and whose direction it is being done.

The fundamental problem is a misalignment between climate foresight and spatial planning. Ecologically sensitive sites, such as former wetlands, flood retention zones, and low-lying areas that served as nature's buffer against intense rainfall and coastal surges, are still being approved for development. These areas, which were once vital for biodiversity and flood mitigation, are now primarily used for private properties, landscaped areas, and paved surfaces with little regard for their hydrological role. Resilience assets are becoming risk multipliers as the region effectively builds over its safety valves.

The fragmented character of development control is one of the main issues facing the corridor. The provincial administrations of Phetchaburi (for Cha-Am) and Prachuap Khiri Khan (for Hua Hin) oversee urban expansion independently, each with its own planning departments, budgetary allotments, and regulatory schedules. Although both jurisdictions are subject to national frameworks like the Disaster Prevention and Mitigation Act and the Town and Country Planning Act, there is no collaboration beyond provincial boundaries. Because of this, integrated floodplain management is rarely accomplished, and infrastructure projects frequently end at jurisdictional borders. Such institutional division is risky and ineffective for a corridor that functions as a de facto single urban region.

Furthermore, there is no single entity in charge of managing large-scale infrastructure investment, climate adaptation, or land-use coherence due to the absence of a corridor-level planning process or metropolitan governance structure. Given the region's rapid growth and climate vulnerability, this lack of strategic control is especially troubling. Development often follows market momentum in the absence of a common spatial vision or uniform zoning system, prioritizing immediate financial gain over long-term resilience.

The situation is made more difficult by the impact of real estate speculation. Due to its closeness to Bangkok, beach attractiveness, and increasing property values, the Hua Hin–Cha-Am corridor has drawn developers, investors, and retirees (Vielhaber et al., 2014). Drainage, access roads, and flood mitigation measures are frequently considered afterthoughts rather than necessities when land is purchased, cleared, and split for houses before any infrastructure is built. As a result, "infrastructure-lite" developments proliferate, which may seem effective during the dry season but fail when exposed to severe weather conditions.

Climate factors are frequently not sufficiently incorporated into regulatory tools, even in officially designed places. When necessary, Environmental Impact Assessments (EIAs) are usually conducted by private consultants, whose objectivity may be questioned, and whose evaluations hardly ever incorporate accurate flood risk modeling under future climate scenarios. There is little guidance on climate-adaptive design, permeable surfaces, green infrastructure, or site elevation standards in flood-prone locations, and master plans and building rules continue to consider flooding as a sporadic annoyance rather than a structural concern.

Additionally, infrastructure development has found it difficult to keep up with growth. Both Hua Hin and Cha-Am have antiquated and disjointed stormwater systems. Many people rely on little concrete drains that were made for rainfall norms from the past rather than current intensities. Large-scale retention ponds are scarce, and when they are, they are frequently neglected or overrun by new construction. In the meantime, major thoroughfares like Phetkasem Road, which are essential for transportation, business, and emergency response, are regularly submerged during intense downpours, highlighting the transport grid's structural vulnerability.

Despite being essential to the corridor's growth, the tourist industry presents a unique set of challenges. Large resort complexes are not always constructed with climate resilience in mind, but rather for beauty and practicality. Few have proper floodproofing or utility system redundancy, despite the fact that many are near the beach, making them more vulnerable to coastal flooding. Access to these resorts is limited, services are interrupted, and insurance premiums increase during floods, resulting in financial strain that is rarely apparent during pre-development planning (see Figs. 8.1, 8.2, 8.3, 8.4, 8.5, 8.6, 8.7, 8.8, 8.9, 8.10, and 8.11).

Vulnerable populations are exposed more because of increased urban pressure. Informal settlements lack basic infrastructure, tenure security, and a say in planning choices. They are frequently found in backlands with poor drainage or on flood-prone borders. Their vulnerable location reflects socioeconomic marginalization:

Fig. 8.1 Agricultural land and rice paddy fields are part of the region's production in seasonal cycle of Thai rice farming. Many examples of post-harvest rice farmlands suggest potential land use changes, land degradation risks, and socio-economic impacts in the regions of Cha-am and Hua Hin

Fig. 8.2 Often in new development corridors like this, growth is associated with land use change

Fig. 8.3 An example of a particular land made ready for new development alongside the new development corridor of Hua Hin-Cha-Am

they bear a disproportionate amount of the burden of flooding while avoiding the advantages of development. In the absence of targeted adaptation investments or inclusive planning processes, these communities are essentially excluded from the corridor's future resilience.

The lack of legally binding coastal setback guidelines is another significant planning shortcoming. Although there are broad guidelines, they are not often followed and are not strictly enforced. Coastal erosion continues to steal property, even as new developments advance closer to the shoreline. Private developers occasionally erect temporary seawalls or obstacles to safeguard their properties, thus shifting the effects of erosion onto nearby coastal areas. The potential for coordinated coastal zone

Fig. 8.4 In such cases, land reclamation and coastal development emerge as widespread practices, often reshaping natural landscapes into agricultural or urbanized zones. These interventions reflect the tension between economic expansion and ecological integrity, where reclaimed fields or modified coastlines become sites of both opportunity and vulnerability

Fig. 8.5 There is always a risk that new development zones will be designated in flood-prone areas, and the corridor's latest planning strategies appear to have overlooked this concern across many kilometres of new development; a long stretch that connects the two struggling provinces

Fig. 8.6 An example of a floodgate near Cha-Am serves as a stark reminder of the severe flooding challenges in the region. While such infrastructure attempts to hold back rising waters, it also symbolizes the growing struggle between rapid development and the realities of a changing climate. The low-lying coastal plains of southeastern Thailand are increasingly vulnerable, not only to seasonal monsoons but also to the intensifying effects of sea-level rise, storm surges, and unpredictable rainfall patterns. Each floodgate marks both resilience and fragility: resilience in the effort to manage water, yet fragility in the recognition that these engineered solutions may only provide temporary relief. As new developments continue to spread across these exposed areas, the question remains whether such measures can withstand the scale of future risks, or whether they merely postpone more difficult conversations about sustainable land use, adaptive planning, and the long-term safety of communities in this corridor

Fig. 8.7 This new development corridor is experiencing more frequent and intense flooding, with recent developments directly exacerbating the problem. Its proximity to large bodies of water, combined with heavy rainfall, leaves the flatlands and plains highly susceptible to recurrent flooding, posing significant risks for long-term development and urbanization strategies. The notion that tourism successes from the western parts of the country can simply be replicated here is concerning, given the stark differences in landform and geographical characteristics. In this corridor, vulnerability is rising at a pace and scale beyond prior expectations. I recall standing on these sites in 2023, warning against the risks of mass development, and by 2025, I found myself witnessing the very projects I had warned about reaching completion

management is compromised by this disorganized collection of individual responses, which also exacerbates cumulative risk.

There is work involved in the region's disjointed planning environment. In order to better control floods, the municipal administrations of Hua Hin and Cha-Am have upgraded specific drainage systems, installed pump stations, and started public awareness campaigns. These initiatives, however, frequently lack the scope, integration, and forward-looking vision required to keep up with the rapid pace of change and are reactive in nature. The areas that will determine the region's course in the ensuing

Fig. 8.8 The thresholds that may no longer be the ones we see today in this new development corridor that connects Cha-Am (in the north) to Hua Hin (in the south). Photo is taken before reaching Hua Hin

decades—land-use regulation, housing policy, and tourism planning—remain largely unintegrated with climate adaptation.

In sum, the Hua Hin–Cha-Am corridor is enmeshed in a planning conundrum: while it develops as a hotspot for development and a destination for lifestyle, its susceptibility to flooding caused by climate change increases with each ineffective response. Systems-level governance—a framework that unifies infrastructure, environment, land use, and social fairness within a climate-resilient paradigm—is what is lacking, not competence or understanding. The seas that currently lap at the borders of new construction may eventually reclaim far more than their share, and development will continue to go against the grain of geography until that time. This chapter's next and last section examines what a more resilient future for the corridor might entail, utilizing both new and unrealized possibilities for coordinated, proactive planning.

Fig. 8.9 Here we see numerous designated sites for new developments—places I passed through while wondering what they might look like by 2035 or 2040, should I have the chance to return and witness their transformation

8.4 Toward Resilient Coastal Futures

The challenge facing planners, policymakers, communities, and investors is not just to reduce risk, but to rethink resilience if the Hua Hin–Cha-Am corridor is a delicate border where rising floods meet rising ambition. Neither development nor decline at any costs are inevitable for this section of the shore. The bravery to design differently, to build with nature rather than against it, and to prioritize long-term sustainability over immediate profits will determine its future.

To achieve this, comprehensive, adaptive, and anticipatory planning must be adopted in place of piecemeal repairs. Realizing that flooding is a structural aspect of living in a changing climate rather than a cyclical threat is the first step towards a resilient coastal future. It needs to be considered from the beginning, not after the fact, in design briefs, investment appraisals, zoning codes, and land-use regulations. Any new road, house, or resort constructed without considering flood risks is a liability in the future. Every paved wetland is a sponge gone.

The Hua Hin–Cha-Am corridor should adopt a new planning paradigm that incorporates the ideas of water-sensitive urban design (WSUD), which views blue-green infrastructure as a fundamental component of urban form and manages rainwater as a resource rather than a nuisance. Permeable pavements, bioswales, green roofs, and retention ponds have to be required for any development permission rather than

Fig. 8.10 The (urbanising) corridor between Cha-Am and Hua Hin is developing far more rapidly than anticipated. Piles of soil and construction machinery now mark landscapes once defined by coconut groves, palm trees, and agricultural plots, signalling the scale of land transformation underway. The steady encroachment of urban expansion is visible not only in new resorts and housing projects but also in the infrastructure of roads, utilities, and power lines that carve across the scene. What is striking here is the layering of change: the natural backdrop of palms remains, but it is increasingly overshadowed by the machinery of growth and the urgency of development. Such rapid construction raises critical questions about land use pressures, environmental vulnerability, and the long-term resilience of this low-lying coastal stretch, particularly as flooding, erosion, and climate risks are intensifying. The scene embodies the paradox of progress—where the promise of economic opportunity collides with the precariousness of ecological balance

Fig. 8.11 There is an increasing demand for new infrastructure and transportation routes in the region. While some projects have proven highly effective along Thailand's southeastern coasts—improving connectivity and supporting local economies—others are placing mounting pressure on fragile land systems and coastal environments. At the same time, several emerging initiatives are being positioned not as isolated projects but as components of a much larger regional development agenda, one that could reshape the broader landscape in the coming years

being optional extras. In addition to being practical, these interventions are also aesthetically pleasing, economical, and advantageous to society.

As natural remedies that protect against surges, lessen erosion, and promote biodiversity, the restoration and conservation of coastal ecosystems—such as mangroves, dunes, and estuary wetlands—should also be given top priority. It is necessary to prioritize soft infrastructure above concrete barriers in low-lying and erosion-prone areas (Allam et al., 2023). Seawalls should be positioned carefully, watched for downstream effects, and combined with environmental buffers where they are judged required. Crucially, it is necessary to enforce coastal setback laws as a moral and ecological requirement rather than as a bureaucratic one.

Reforming governance is equally important. The corridor's ability to design and function as a single socio-ecological system is weakened by the current administrative separation between Hua Hin and Cha-Am. A corridor-level coordination framework that can manage land use, climate adaptation, infrastructure investment, and disaster response in a coordinated way is required (Cheshmehzangi, 2025); this might be an interprovincial planning board or coastal climate taskforce. Threats from climate change transcend national boundaries, and our resilience plans should too.

Climate-responsive urbanism must become the norm, not the exception. At the same time, current urban masterplans need to be reviewed and updated to include climate projections and hydrological modelling, which means modifying land-use allocations based on flood risk maps, raising critical infrastructure above projected flood levels, and prohibiting construction in known floodplains unless adaptive design solutions are used. Additionally, building codes need to be improved, and developers should be encouraged to use elevated foundations, raised utilities, and site-specific water retention features.

However, without local people' involvement and empowerment, regulations and infrastructure alone are insufficient. Throughout the corridor, community-based adaptation that is based on equitable benefit sharing, participatory planning, and local knowledge should be encouraged and scaled. This includes flood preparedness education, early warning systems, and retrofitting initiatives at the home level. In both pre-disaster planning and post-disaster recovery, local voices—particularly those in underserved or informal areas—must be heard. After all, resilience refers to people's ability to adjust, react, and prosper rather than just infrastructure.

As the region's main source of income, tourism has the potential to foster resilience. Climate-conscious travel destinations, locations that offer comfort and beauty while also showcasing sustainability in action are becoming increasingly popular worldwide. Flood-resilient public areas, eco-resorts, and green-certified construction could make Hua Hin–Cha-Am a shining example of coastal innovation rather than a warning. With hotel chains and real estate developers funding inclusive resilience projects, coastal protection, and local adaption infrastructure, public–private partnerships might have a revolutionary impact in this area.

The most significant change in the future might be conceptual: a change from fighting water to embracing it. This entails creating cities with raised boardwalks, floating marketplaces, adaptive landscapes, and amphibious structures that can adapt to seasonal flooding rather than just fend it off. It entails envisioning a corridor that flows with the tide rather than against it—a location where development occurs in harmony with nature rather than in opposition to it.

In conclusion, the coastline corridor between Hua Hin and Cha-Am is at a turning point. Decades will be affected by the decisions taken in the upcoming years. If the current course, i.e., piecemeal, reactive, and uncoordinated, continues, the area would be caught in a vicious cycle of costly repairs and recurrent disasters. If a new approach is taken—one that is proactive, inclusive, and ecologically based—this corridor has the potential to emerge as a model of resilient coastal urbanism in Thailand and Southeast Asia as a whole. There is yet time. Our collective imagination can rise together with the seas. There is yet room to construct something more intelligent, robust, and long-lasting between the sea and sprawl.

References

Allam, Z., Cheshmehzangi, A., & Jones, D. S. (2023). The economic landscape of cities and climate change. In *Climate and social justice: The political economy of urban resilience and mercantilism* (pp. 25–41). Springer Nature Singapore.

Arunothaipipat, R. (2011). Human resource management of small independent hotels: A case study of Cha-am Beach, Phetchaburi, Thailand. *SIU Journal of Management*, 88–100. https://www.academia.edu/download/32127312/rosamarin.pdf

Baye, F. (2025). Exploring urban infrastructure challenges in informal Peri-Urban Woldia: Barriers, implications, and informal strategies. *Frontiers in Sustainable Resource Management, 4*, 1555564.

Cheshmehzangi, A. (2025). The importance of health-sustainability-peace nexus in making resilient cities. *Peace and Sustainability, 1*(2), Article 100012.

Cheshmehzangi, A., Sedrez, M., & Flynn, A. (2024). Stormwater management: Issues, opportunities, and challenges in cities and communities. In *Rethinking stormwater management through sustainable urban design* (pp. 1–22). Springer.

Husa, K., Vielhaber, C., Jöstl, J., Veress, K., & Wieser, B. (2014). Searching for paradise? International retirement migration to Thailand—A case study of Hua Hin and Cha-am. In *Southeast Asian mobility transitions: Issues and trends in migration and tourism*, 137–167.

Kityuttachai, K., Tripathi, N. K., Tipdecho, T., & Shrestha, R. (2013). CA-Markov analysis of constrained coastal urban growth modeling: Hua Hin seaside city, Thailand. *Sustainability, 5*(4), 1480–1500.

Sealeum, S., Chankaew, K., Inkapattanakul, W., & Navanugraha, C. (2007). *Environmental quality assessment criteria for recreational beach: A case study of Cha-am beach, Petchburi Province.* https://www.thaiscience.info/Journals/Article/ENRJ/10892395.pdf

Tanaboriboon, T. (2022). The performance motivation of Cha-am municipality officers Phetchaburi Province. *International Journal of Development Administration Research, 5*(2), 1–8.

Thanarotrungrueang, K., Uamsiri, S., Talalak, K., & Tubbueng, T. (2022). Chinese-English mobile applications for communication of heal massage personnel to enhance employment opportunities: Case study of Cha-Am and Hua-Hin areas. *Life Sciences and Environment Journal, 23*(2), 437–450.

Unhasuta, S., Sasaki, N., & Kim, S. M. (2021). Impacts of tourism development on coastal communities in Cha-am Beach, the Gulf of Thailand, through analysis of local perceptions. *Sustainability, 13*(8), 4423.

Vielhaber, C., Husa, K., Jöstl, J., Veress, K., & Wieser, B. (2014). Paradise found? Experiences of Farang retirement migrants in Hua Hin and Cha–am, Thailand. In *Southeast Asian mobility transitions: Issues and trends in Migration and Tourism* (pp. 168–195).

Chapter 9
Three Beaches, One Challenge: Integrated Coastal Management for Climate Resilience in Bangtao, Karon, and Rawai of Phuket Island

9.1 Setting the Scene: Phuket's Vulnerable Shorelines

The largest island in Thailand, Phuket, has long been praised for its azure waters, fine sands, and thriving coastal communities. However, underneath this idyllic picture lurks a reality of increasing environmental susceptibility. Phuket is a major tourist destination, and its beaches serve as both leisure areas and cultural landmarks. Because of their dual function, it is both ecologically and socioeconomically imperative that they be preserved.

Three separate coastal zones, Bangtao, Karon, and Rawai, represent a variety of Phuket's coastal management issues. These are not connected but are selected due to their various levels of tourism activities and development in the region, relatively lower than some other coastal zones in Phuket Island. Every beach exhibits a distinct combination of environmental stresses, community structure, development trends, and terrain. Their adaptive responses differ, influenced by local histories, governance structures, and resource limitations, but sharing the same maritime and climatic stressors.

The main issue these beaches face is the growing combination of pressures caused by human activities and natural risks. Low-lying areas frequently experience flooding due to storm surges that push seawater far inland during seasonal monsoon rains. Natural barriers like dunes and mangroves have been weakened by coastal erosion (Cheshmehzangi, 2025a), which has been made worse by uncontrolled building and sand mining in recent decades. In the meantime, the dangers are increased by the increasing occurrence and unpredictability of extreme weather events associated with climate change.

The island's economy depends on tourism, which can both increase risk and foster resilience. The quick growth of beachside infrastructure, such as lodging facilities, dining establishments, and business centers, has made areas more vulnerable to storm damage and decreased natural ecosystems' ability to adapt. However, the sector's

A. Cheshmehzangi, *Thailand's Tropical Thresholds*, Urban Sustainability,
https://doi.org/10.1007/978-981-95-4975-7_9

economic clout also gives investors power to fund preventive measures like ecosystem restoration and designed seawalls.

The intricate interaction between national frameworks, provincial mandates, and local efforts is reflected in Phuket's coastline government. Broad resilience priorities are determined by national policies like Thailand's Coastal Resources Management Plan, but local implementation is contingent upon the ability and willingness of private stakeholders, village leaders, and municipal authorities. The gaps left by scarce public resources are frequently filled by community-driven activity; in the case of beach clean-ups, mangrove regeneration, and disaster preparedness, grassroots initiatives are crucial.

One example is Bangtao Beach, which is located on the west shore of the island. Its lengthy, gently sloping coastline is susceptible to both storm-driven waves and seasonal flooding from inland drainage. Luxury resort construction has exploded in the area, increasing land value but complicating land use planning. Further south, Karon Beach is increasingly urbanized and heavily dependent on large-scale tourism (Wongwattanakit et al., 2017). During the rainy season, it has ongoing drainage issues, and both businesses and inhabitants are impacted by flash floods (Otto, 2016). On the southeast tip, Rawai Beach is more directly related to traditional coastal subsistence and fishing livelihoods. Its lower altitude and vulnerability to storm surges endanger not just tourist attractions but also locals' houses and means of subsistence.

Notwithstanding their variations, all three beaches urgently require integrated coastal management—methods that strike a balance between the maintenance of natural systems and constructed defences, addressing both short-term risks and long-term climate resilience. They also highlight the value of comparative analysis, which can be used to learn how to scale up effective approaches and steer clear of expensive blunders by analyzing where techniques overlap or differ.

Living along the edge—between land and sea, between development and conservation, between immediate profit and long-term sustainability—is at the heart of Phuket's waterfront tale. Bangtao, Karon, and Rawai provide as living examples of how small coastal economies may preserve their natural and cultural history while adjusting to increasing environmental constraints. In order to better understand these beaches' distinct problems, management strategies, and prospects for a more resilient future (Cheshmehzangi, 2025b), this chapter will compare and contrast them.

9.1.1 Comparative Context of Bangtao, Karon, and Rawai, Phuket Island

In addition to being geographical sites, the three beaches at the heart of this study—Bangtao, Karon, and Rawai—are also unique socio-environmental systems. Interpreting the achievements and failures of various coastal management initiatives requires an understanding of their comparative background.

Bangtao Beach is one of Phuket's longest beaches, spanning more than six kilometers. Lagoons and mangroves border its northern portions, which historically provide substantial natural flood control. Both economic growth and environmental stress have resulted from the area's slow transition into a posh resort destination, especially in the vicinity of the Laguna Phuket complex. Although tourism-related jobs dominate the local economy, some residents are still employed in inland fishing and small-scale farming. During the southwest monsoon, when runoff from nearby hills overwhelms drainage systems, seasonal flooding is frequent. Luxury resorts have made investments in beach nourishment and private flood defences, but public infrastructure has not kept up, resulting in inconsistent protection throughout the region.

South of Patong, Karon Beach is a thriving urbanized strip that is well-known for its busy nightlife, large number of beachfront hotels, and high volume of visitors. Its expansive sandy area is a strength and a weakness; although it offers recreational space, its openness to wind and waves leaves it exposed to storm damage. Karon has long-standing drainage problems; flash flooding frequently occurs in low-lying commercial areas due to clogged or inadequate culverts (Kessaratikoon et al., 2013). Since a large portion of Karon has been destroyed for urban development, the area has less mangrove or wetland buffers than Bangtao. Nonetheless, the municipality has taken the initiative to implement beach clean-up efforts, public awareness campaigns, and small-scale drainage improvements.

Rawai Beach, at the southern end of the island, presents a markedly different character. It is more about livelihoods than it is about sunbathing. Deeply ingrained ties to the sea are reflected in the cultural environment, bustling seafood markets, and fishing boats lining the shore. Although Rawai is a gateway for marine tourism due to its closeness to offshore islands, its infrastructure is still lacking in comparison to the west coast. Residents and businesses depend on the low-lying coastal route, which is frequently submerged during storm surges. Flooding here has an effect on both tourism and the fishing business, as lost catch, broken gear, and interrupted transportation pose major financial risks. Local leaders have been outspoken supporters of mangrove restoration and the incorporation of traditional ecological knowledge into official planning procedures, and community networks are robust.

Comparative observations highlight both common weaknesses and unique strengths. Although Bangtao's upscale tourism industry makes it easier for the private sector to invest in resilience measures, it also makes the city more dependent on one industry. Karon's high-density growth makes systemic drainage upgrades essential (Pongprayoon, 2006), because it exacerbates the effects of infrastructure failure. With its mixed economy of fishing and tourism, Rawai offers a wider range of economic options, but it is also more vulnerable to environmental change since it lacks the financial protection of luxury travel.

Baselines for the environment also vary. Mangroves and lagoons, two of Bangtao's remaining natural buffers, could improve flood protection if they are preserved and rehabilitated. Despite lacking these characteristics, Karon has a rather broad beach profile that helps disperse wave force. The seagrass beds and mangroves of Rawai are vital habitats that serve as barriers of protection, but they are being strained by overuse and development.

Though the proportion of state-led to community-led action varies, all three beaches function within Phuket's larger municipal and provincial institutions from a governance standpoint. Resort developers in Bangtao frequently establish the objectives for coastal management, sometimes in accordance with state policy and other times not. Karon's strategy emphasizes formal infrastructure planning and is better integrated at the local level (Pongprayoon, 2006). In contrast, Rawai places a strong emphasis on grassroots initiatives, with local leaders and NGOs working with provincial agencies to shape adaptation strategies.

This comparative perspective emphasizes the necessity of an integrated strategy that combines the adaptive ingenuity of fishing communities, the financial ability of tourism zones, and the planning discipline of municipal authorities. Understanding these interdependencies and encouraging cooperation between beaches, sectors, and governmental levels will be essential to Phuket's coastline's resiliency. By understanding the context of Bangtao, Karon, and Rawai side by side, we can better identify which strategies are transferable, which are site-specific, and how to bridge the gaps that leave certain areas more vulnerable than others.

9.2 Coastal Hazards and Risk Profiles

The shoreline of Phuket is a dynamic border that has been molded by silt, wind, ocean, and human activity. This dynamic is most noticeable in Bangtao, Karon, and Rawai, three beaches that represent distinct manifestations of hazard exposure despite their close proximity. Stronger monsoons, erratic storm surges, increasing sea levels, and erosion made worse by unchecked development are just a few of the environmental threats that each faces as a result of climate change. However, the form and degree of these risks vary, creating unique risk profiles that need to be recognized before implementing any successful management plan.

9.2.1 Monsoon Flooding and Inland Runoff

Heavy rainfall brought on by the southwest monsoon, which usually lasts from May to October, can overflow drainage systems, flood low-lying areas, and result in interior flooding that eventually flows into the beaches.

Bangtao: Several small rivers and canals, as well as mountainous terrain, make up the catchment area behind Bangtao. Runoff speeds down these slopes during heavy rainstorms, gathering silt and debris before dumping into the lagoons and the shore. Land reclamation and infilling have decreased the lagoons' storage capacity, despite the fact that they once served as buffers, absorbing surplus water (Antoh, 2021).

Floods occur more frequently in residential and commercial areas because drainage routes are frequently obstructed by silt or encroachment.

Karon: Due to the city's crowded metropolitan area, moving water fast is more of an issue than the water's source. Within hours of a downpour, streets can turn into rivers because rainwater is trapped by impermeable surfaces like parking lots, pavement, and roadways. Undersized stormwater pipes and a lack of culverts limit the outflow, resulting in standing water for extended periods of time that harms infrastructure, interferes with tourism, and increases public health hazards like mosquito breeding.

Rawai: Tidal cycles often exacerbate flooding in Rawai. Stormwater from interior towns flows toward the coast during monsoon rains, but if significant rain falls during high tide, water backs up into fishing settlements and low-lying highways. Seawater can occasionally push its way into the drainage system, polluting freshwater sources and hastening the degradation of road surfaces.

9.2.2 *Storm Surges and Extreme Weather Events*

Although the 2004 Indian Ocean tsunami (Choowong et al., 2008) serves as a vivid reminder of Phuket's susceptibility to sporadic yet devastating occurrences, storm surges linked to tropical depressions and typhoons that erupt in the Andaman Sea or elsewhere provide a more frequent hazard (Wongbusarakum, 2002).

Bangtao: During a surge, wave energy can flow well inland due to the wide, gently sloping beach profile, especially when strong onshore winds are present. When there are no ongoing dune systems, constructed infrastructure is left open.

Karon: Due to its open southwest orientation, which permits storm-driven waves to strike the coast with little hindrance, Karon is particularly vulnerable. Direct effects include structural damage and seawater intrusion into foundations for hotels and businesses situated near the high tide line.

Rawai: Despite being somewhat protected by offshore islands, Rawai is vulnerable to water overflowing the barrier or coastal road due to its lower elevation (Buakhao & Weerakit, 2013). Particularly at risk are fishing boats anchored along the shore; powerful surges have the potential to break mooring lines and destroy gear, threatening local livelihoods.

Large amounts of debris, sand, and even coral rubble have been reported to be deposited inland by storm surges in all three areas, necessitating expensive cleanup and the temporary closure of tourist attractions.

9.2.3 *Coastal Erosion*

Although erosion is a natural process, human activities like sand mining, shoreline development, and the removal of protecting vegetation have expedited it in Phuket.

Stronger wave action and changed sediment transport patterns brought on by climate change exacerbate these demands.

Bangtao: Erosion is especially noticeable in areas where native vegetation has been removed to make way for construction. Sand is only partially replenished during the dry season after being removed by seasonal currents during the monsoon (Martin, 2010, 2020). In certain cases, seawalls that have been put in place deflect wave energy downward, further eroding the coastline.

Karon: Although the width of this vast sandy stretch varies seasonally, there has been a net loss of beach area in recent years. If drainage outlets are not kept up, they may form rills that allow runoff to pass through the sand and compromise the stability of the beach. Beach replenishment has been tried occasionally, but results are short-lived in the absence of comprehensive sediment control.

Rawai: Mangrove deterioration and erosion interact here. The shoreline recedes more swiftly in places where mangroves have been removed for aquaculture or construction, leaving buildings and roads vulnerable to direct wave onslaught. Additionally, the removal of natural barriers lowers the capacity for carbon sequestration and biodiversity.

9.2.4 Sea-Level Rise and Salinisation

Sea levels in the Andaman region may increase by 0.4 to 0.8 m by the end of the century under high-emissions scenarios, according to projections from the Intergovernmental Panel on Climate Change (IPCC). In low-lying coastal areas, even slight elevation changes have significant effects.

Some parts of Bangtao have reported saltwater intrusion into groundwater, which has an impact on the availability of freshwater for locals and tourist attractions. High-value coastal residences in Karon are at risk of both immediate flooding and the slow deterioration of septic systems, which could compromise the quality of the water. Last but not least, coastal agricultural and aquaculture ponds in Rawai are most severely affected, as rising salinity can lower yields or necessitate adjustments to production systems.

9.2.5 Ecosystem Degradation and Reduced Natural Defences

Mangroves, coral reefs, seagrass beds, and sand dunes all serve as vital natural defences, attenuating wave energy and stabilising shorelines. However, both direct exploitation and indirect pressures have gradually weakened these systems throughout Phuket. Widespread bleaching caused by rising waters has affected coral reefs, which formerly assisted in releasing wave energy before it reached the coast. Regular boat anchoring and sedimentation from inland drainage have harmed the seagrass habitats close to Rawai and Bangtao. Once prevalent in these same locations, mangroves have lost some of their ability to serve as natural flood barriers due to their clearing for aquaculture and resort development. In addition to weakening coastal resilience, the loss of these interdependent ecosystems makes people rely more on manmade structures, which are frequently more costly to maintain and much less flexible in the face of climate change.

9.2.6 *Human-Induced Exacerbators of Risk*

In addition to the effects of nature, human choices have significantly increased Phuket's coastal regions' susceptibility and fragility. Communities are now forced to rely on hard defences like seawalls, which over time can worsen erosion, because coastal construction that presses too near to the high tide line has left little area for natural beach migration. Seasonal downpours become devastating floods that persist long after the rains have finished due to poorly planned or obstructed drainage pipes.

Tourism pressures add another layer of strain, as high visitor numbers overwhelm waste management systems and contribute to water pollution that weakens already stressed coastal ecosystems. By eliminating essential beach sediment in previous decades, uncontrolled sand extraction further weakened the island's natural defences, making certain areas permanently narrower and more vulnerable to wave assault. In addition to weakening natural systems' resilience, these human-caused stresses have solidified a cycle of reliance on expensive and less adaptable manmade remedies (Fig. 9.1).

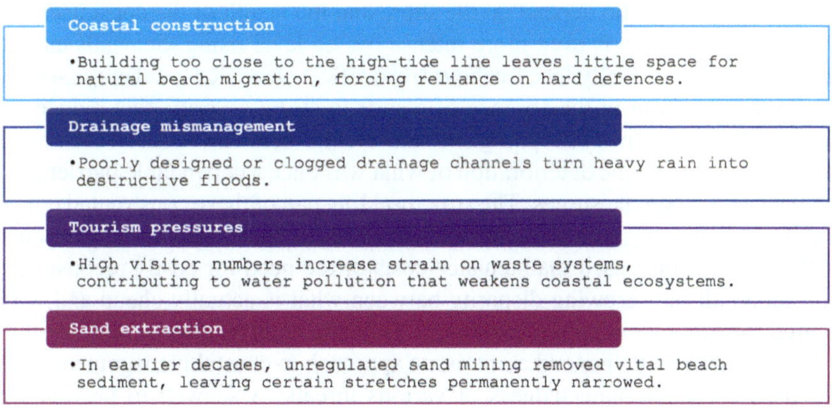

Fig. 9.1 Summary of key human-causes stresses and challenges in the region

9.2.7 Combined and Cascading Risks

The possibility of cascading impacts—circumstances in which one hazard exacerbates the consequences of another, resulting in a series of events that increase overall risk—is one of the most important and frequently overlooked aspects of Phuket's coastal hazard profile. These interactions have the potential to transform an otherwise controllable hazard into a protracted and destructive crisis in a dynamic coastal environment such as Phuket's (Kiriwongwattana & Waiyasusri, 2024). For example, even well-designed defences may be overwhelmed by a storm surge that occurs during the monsoon season at high tide, sending seawater well inland. The issue is even worse when such an occurrence coincides with a period of intense rainfall since floodwaters have nowhere to go. Water is further trapped by clogged or insufficient drainage systems, leading to extended flooding that damages infrastructure, disturbs livelihoods, and raises the risk of waterborne illnesses (Cheshmehzangi et al., 2025). These interrelated factors produce a setting where damage costs sharply increase, and recovery is slower.

Cascading impacts are not limited to short-term weather events; they can also unfold through slower, more insidious processes that steadily erode a coastline's resilience. For instance, erosion is more than just the disappearance of sand from a beach; it also involves the loss of vital vegetation that serves as a natural barrier against wave energy, including dune grasses and mangroves. The shoreline is far more vulnerable to future storm surges once these vegetative defences are removed. This accelerates the rate of coastal erosion and deprives communities of their natural safeguards by creating a feedback loop where each storm or high-wave occurrence causes more damage than the one before it. Ecological repercussions of vegetation loss include decreased biodiversity and weakened natural systems that support the lives of individuals who depend on the shore and the health of the coast itself.

This chain reaction is even more intense by sea level rising. Seawalls and other coastal defences gradually lose freeboard, or vertical clearance, as average sea levels rise. Over time, even in the absence of a shift in storm severity, this reduces their efficacy. More frequent overtopping and seawater intrusion into coastal communities are made possible by the deterioration of what was once a dependable barrier against high tides and moderate surges. The structural lifespan of these constructed defences is limited when rising seas are coupled with stronger storms and more wave energy, which results in increased maintenance costs or the need for costly reconstruction. This can result in a growing disparity between what is actually cheap and what is required to keep safe in communities with little financial means.

The necessity to go beyond a one-issue approach to coastal management is highlighted by the interconnectedness of various threats. A seawall to prevent storm surges is an example of an intervention that is too frequently created to address a single hazard without properly considering how that measure could interact with other dangers or even generate new vulnerabilities. Poorly designed flood-control systems can trap rainfall inland when high tides prohibit it from being released to the sea, and a seawall that impedes natural sand movement can exacerbate erosion downstream (Neef et al., 2018). This disjointed strategy runs the risk of producing the exact kind of cascade failures that expose communities more rather than less.

In Phuket, where fishing, tourism, and coastal communities are closely linked, cascading effects have both social and economic repercussions. Cultural landmarks may be lost to the water, businesses may experience frequent closures, and surrounding people may witness a drop in property values. The interdependence of livelihoods means that damage in one sector can ripple outward—for example, when storm damage to fishing infrastructure limits seafood supply, affecting restaurants and market vendors who in turn rely on tourism income. It takes more time, money, and coordination to recover from such compounded crises, which puts more strain on private players, community organizations, and municipal governments.

Ultimately, identifying and preparing for cascading effects is crucial to constructing climate resilience and goes beyond simple academic study. Coastal managers can create integrated adaptation plans that stop minor issues from turning into major catastrophes by modeling and predicting how hazards may overlap or interact. This could entail establishing development restrictions in areas where several risks are anticipated to converge, constructing drainage systems that continue to operate under tidal pressure, or combining engineered solutions with the restoration of natural buffers. The resilience of Phuket's coastline will depend on the capacity to act pro-actively, think holistically, and manage the coast as a complex, linked system rather than a collection of discrete hazard zones, as demonstrated by the experiences of Bangtao, Karon, and Rawai.

9.2.8 Risk Perception and Adaptive Capacity

Local perceptions of risk influence how communities plan and react, even while technical assessments offer useful information about dangers.

High-end resorts in Bangtao frequently make significant investments in security measures, although residential neighbourhoods and smaller companies do not have the same funding. Even though both are essential, the municipality of Karon occasionally prioritizes sustaining tourism activities over long-term ecosystem rehabilitation. Strong community networks in Rawai enable quick informal responses to emergencies, but preventive actions are hindered by a lack of financing.

The three beaches differ greatly in their adaptive capacity, which is the power to respond to consequences, seize opportunities, and adapt to possible harm. Effective governance, financial resources, social cohesiveness, and the state of ecological systems all have an impact.

In conclusion, the risk and hazard profiles of Bangtao, Karon, and Rawai, all point to the same reality: both man-made and natural hazards are growing more severe. The challenge's physical core is made up of monsoon flooding, storm surges, erosion, and sea level rise, but its underlying vulnerabilities are caused by ecosystem degradation, poor infrastructure, and occasionally misdirected development priorities. Every beach has a unique set of hazards, yet because Phuket's coastline is interconnected, no place can afford to take isolated action. The following sections will look at how localized management and adaptation techniques are being used to address these risks, as well as how lessons may be learned for integrated resilience strategies that apply to the entire island.

9.3 Localised Management and Adaptation Measures

Efforts to control coastal hazards and increase resilience in Bangtao, Karon, and Rawai combine community-led projects, nature-based treatments, engineered solutions, and regulatory frameworks. These actions vary in scope and level of complexity, but taken as a whole, they serve as the foundation for any beach's adaptation plan. While some result from urgent reactions to current challenges, others are the result of long-term planning and policy direction. When taken as a whole, they show how intricate and multi-layered coastal management is in an environment where livelihoods, tourism, and environmental integrity are all under risk.

The most obvious manifestation of localized control is still seen in engineered infrastructure, especially in the form of retention ponds, drainage systems, and seawalls. Seawalls that are intended to withstand storm surges and save priceless resort houses have been installed along portions of Bangtao's shoreline (see Figs. 9.2, 9.3, 9.4, 9.5, 9.6, 9.7, and 9.8). Although these buildings offer instant protection, they are costly to maintain and, by interfering with the normal flow of sand, can occasionally worsen erosion in nearby areas. Enhancing the drainage system has been the main goal in Karon to combat persistent floods during the monsoon season. The goal of new culverts and improved stormwater pipes is to transport water from streets to the sea more effectively, but their performance is still hindered when high tides and heavy rains occur at the same time. In order to lower the risk of inland flooding, Rawai has made investments in smaller-scale interventions such maintaining retention ponds that temporarily store runoff and improving the drainage outlets on the coastal road. Although these engineering procedures are crucial for immediate hazard mitigation, they can only provide partial and frequently transitory answers in the absence of supporting efforts.

The limitations of hard infrastructure have led to a greater focus on ecosystem-based strategies that complement natural processes rather than interfere with them. Mangrove restoration initiatives are reestablishing a first line of defence against storm surges and waves in Rawai and northern Bangtao. Replanting native species, enhancing tidal flows, and stopping more encroachment are all part of these initiatives, which are frequently spearheaded by collaborations between local government agencies, non-governmental organizations, and community organizations. In other areas of Bangtao and Karon, dune restoration has also gained popularity. There, sand fencing and vegetation planting assist stabilize the beach and lessen erosion caused by wind. Artificial reef deployment off the coast is being utilized to break wave energy before it reaches the beach, in addition to supporting marine wildlife. Once developed, these nature-based initiatives typically require less intense maintenance,

Fig. 9.2 Bangtao shorelines are filled with many high-end housing and resorts indicating massive tourism-based changes in the region

Fig. 9.3 Coastal resorts and tourism shape most of Bangtao's shoreline, where local character and coastal charm intersect with the pressures of tourism-driven transformation

Fig. 9.4 Strained infrastructure and drainage systems are mixed with the everyday realities of mobility in Bangtao

Fig. 9.5 Local communities are more vulnerable than what meets the eye in Bangtao

Fig. 9.6 Local communities also suffer from inadequate water systems and clean water supply

offer co-benefits for tourism and fisheries, and are more adaptive than hard constructions. However, to guarantee long-term success, they require constant protection and observation.

Hazard setbacks, construction codes, and land use zoning are additional crucial pillars of localized adaptation. Building setback regulations have been put in place by Karon's municipal officials in an effort to limit exposure to and strain on protecting ecosystems by preventing new construction from coming too near the high tide line. Although the demands of high-end resort development occasionally compromise Bangtao's zoning rules, they do make an effort to strike a balance between the expansion of tourism and environmental constraints by establishing no-build zones in regions that are especially sensitive. Although less codified, Rawai's policies are developing to incorporate hazard mapping into planning choices, guaranteeing that high-density development does not occur in sensitive areas, particularly those susceptible to tidal flooding. Enforcement of these laws is crucial to their efficacy, but it can be uneven when powerful business interests are involved. However, they are crucial for encouraging more proactive, risk-sensitive growth and ending the cycle of reactive infrastructure construction.

Initiatives for disaster preparedness and community awareness provide these adaptation efforts a crucial social component. Public education initiatives have aimed to raise awareness of climate-related risks and promote proactive engagement in resilience-building at all three beaches. Seasonal workshops on storm preparation, safe mooring techniques, and evacuation planning are held in Rawai by village

Fig. 9.7 Bangtao's shoreline reminds me of a question I had in mind when I first visited the Maldives in 2007, that why the visitors get to enjoy the coasts much more than locals? And why the over-commercialisation of coastal areas is a common practice in developing nations in compared to the so-called developed ones?

committees and fishing cooperatives. Schools in Karon include disaster awareness in their curricula to make sure that the next generation is raised knowing the fundamentals of local hazards and how to respond to them. With hotels and dive shops organizing beach clean-ups, mangrove planting days, and visitor education sessions on environmental stewardship, Bangtao's community involvement has frequently been connected to tourism businesses. These programs increase social capital, which in times of crisis can be just as crucial as physical infrastructure.

Crucially, integrating these various strategies is frequently essential to the effectiveness of localized management initiatives. Community preparedness without supporting infrastructure may lessen immediate harm but cannot stop long-term damage; a seawall without drainage upgrades may cause backflow issues during periods of heavy rain; and mangrove regeneration without zoning protection may be jeopardized by adjacent development. For instance, in Bangtao, the most resilient beach stretches are those where resorts have integrated dune restoration, engineered barriers, and proactive community engagement. In Karon, better drainage has worked best when it coexists with low-lying, flood-prone areas' ban on high-density construction. When combined with fishing community involvement, Rawai's continuous mangrove restoration has a bigger impact and keeps local livelihoods in line with environmental preservation.

Fig. 9.8 Like many others, the extended beaches of Bangtao are filled with tourism activities, sport activities, and many commercial offerings and businesses

Thus, it is easiest to think of localized adaptation in Phuket's coastal zones as a patchwork of strategies, each of which addresses a different aspect of the threat. Some solutions try to foresee and lower future risks, while others directly address past calamities. Each beach's overall resilience is determined by the interaction of hard defences, nature-based solutions, regulatory constraints, and community action. Strengthening these connections, scaling up effective models, and making sure that temporary safeguards are included into long-term plans for sustainable coastal living will be the challenges of the future. In the context of Bangtao, Karon, and Rawai, the lessons are clear: resilience is most robust when it draws on multiple, complementary approaches that balance immediate protection with the restoration and preservation of the very ecosystems that make life on the coast possible.

9.4 Comparative Analysis of Approaches and Effectiveness

The coastline management strategies of Bangtao, Karon, and Rawai, which are influenced by topography, economic forces, governmental frameworks, and community interests, exhibit both obvious differences and commonalities when compared side by side. Every beach has made investments in a variety of ecosystem-based, engineering, and regulatory measures, but the proportions of each method vary, which affects resilience as a whole.

Because of its upscale resort economy, Bangtao enjoys substantial private sector investment. This has made it possible to invest heavily on defensive infrastructure such private drainage systems, beach replenishment, and seawalls. To provide a layered defence, these actions are sometimes combined with mangrove rehabilitation and dune restoration. However, because of the dependence on privately run enterprises, protection may not be uniform, with less focus being placed on areas outside of tourist zones. Furthermore, development may be pushed closer to the shoreline by commercial demands, which compromises zoning intentions and strains natural buffers. Bangtao shows the dangers of fragmented protection when state monitoring is inadequate, but it also shows how private capital may speed up adaptation.

As one of Phuket's most popular tourist destinations, Karon takes a more municipally controlled approach. In order to solve the ongoing flooding that causes business disruptions throughout the monsoon season, drainage improvements have been prioritized here. Additionally, the town has put in place building setback rules, which, when followed, aid in maintaining a certain amount of natural beach area. However, chances for extensive ecosystem restoration are limited by Karon's highly urbanized nature. Although small-scale trials of vegetation planting and artificial dune construction have been conducted, their effects pale in comparison to those of hard infrastructure. The end effect is a system that does a fair job of controlling seasonal flooding, but it is still quite vulnerable to erosion and storm surges, especially in areas without protecting vegetation (see Figs. 9.9, 9.10, 9.11, 9.12, 9.13, 9.14, and 9.15).

Rawai is unique in that it effectively combines nature-based solutions with community-led action. In addition to restoring important natural barriers, mangrove rehabilitation programs led by regional fishing cooperatives and non-governmental organizations have strengthened the economic and cultural ties between locals and the sea. Smaller-scale drainage upgrades and hazard mapping programs that support local decision-making are added to these projects. However, compared to Bangtao or Karon, Rawai has significantly fewer financial resources, which limits its ability to modernize its infrastructure on a broad scale. This indicates that without extra engineered reinforcement, its ecosystem-based strategy may be less effective in the face of abrupt, high-intensity storm events, even while it offers significant long-term resilience benefits.

Comparatively speaking, the most robust results occur when several strategies are combined and reinforce one another. Sections of Bangtao that integrate seawalls with dune restoration and community involvement are better equipped to endure seasonal dangers than those that only use hard structures. In Karon, land-use restrictions that prohibit high-density construction in flood-prone regions are most effective when combined with drainage improvements. Mangrove restoration in Rawai is more effective when combined with focused infrastructure upgrades that safeguard fishing grounds and susceptible access routes.

Fig. 9.9 Designated Tsunami hazard zone signs in coastal areas of Karon speak very loudly for themselves, indicating the seriousness of flooding-related issues in the area

The distinctions between the three beaches also emphasize how important stakeholder collaboration and governance are. Because Bangtao depends on private investment, more robust governmental monitoring is necessary to provide fair protection. Although Karon's municipal leadership offers some consistency, it has to figure out how to include more ecosystem-based strategies. To close the gap between community initiative and extensive hazard reduction, Rawai's grassroots strength requires financial and technical support.

Finally, the comparative study indicates that no one tactic is enough on its own. Combining hard and soft measures, coordinating actions across sectors and scales, and coordinating public and private agendas are all necessary for resilience in Phuket's coastal zones. Bangtao, Karon, and Rawai may work toward a more cohesive, island-wide strategy that safeguards not only their shorelines but also the

Fig. 9.10 Hard and soft landscapes are mixed here. The pandanus trees and other vegetations are part of the shorelines of Karon in any way possible

communities and economy that rely on them by sharing knowledge about each other's achievements and failures.

9.4.1 Lessons from Similarities and Differences

Despite having different physical locations and socioeconomic backgrounds, Bangtao, Karon, and Rawai all face similar difficulties that unite them into a single story of adaptation in Phuket. The most valuable lessons for improving climate resilience and coastal management on the island, however, come from their

Fig. 9.11 Karon's coastal management suffers from excessive tourism activities and resorts at its shores

Fig. 9.12 Meanwhile, Karon benefits from vast area of vegetation in compared to its nearby coastal areas

Fig. 9.13 A warning flag and storm-displaced debris on Karon Beach highlight the vulnerabilities of this coastline to flooding and high tides. Seasonal surges and shifting tidal patterns often wash heavy objects ashore, leaving behind signs of the ocean's force and the growing challenge of managing coastal safety in Phuket's popular tourist zones

Fig. 9.14 Coastal pandanus trees along the beach in Karon provide vital shade and natural protection, their dense aerial roots anchoring the sandy soil against erosion. These resilient plants embody both ecological value and cultural presence, offering spaces of rest while buffering the shoreline from the impacts of wind, waves, and rising seas

differences—their disparate approaches to governance, infrastructure, ecosystem protection, and community engagement.

Their mutual vulnerability to a mix of storm surges, seasonal floods, and increasing erosion is one of the most obvious parallels. The high rainfall of the southwest monsoon puts all three beaches' drainage systems and coastal buffers to the strain. Sea level rise will gradually reduce the efficiency of current protective features, such as mangroves, dunes, and seawalls, in the future. The necessity of integrated, island-wide design that views the shoreline as a single, continuous system rather than a collection of discrete sections is highlighted by this shared danger profile. Piecemeal interventions can have unexpected effects elsewhere since the hydrodynamic and sediment processes at one place always affect conditions further along the shore.

However, the variations in reaction mechanisms and local capabilities are equally illuminating. Large-scale infrastructure can give immediate protection benefits thanks to private investment made possible by Bangtao's resort-driven economy. However, this paradigm runs the risk of producing uneven resilience, with nearby community areas remaining under-protected while high-value homes enjoy significant defences. The takeaway from this is that private capital can be an effective adaptation tool—but only if it is used in conjunction with robust regulatory frameworks and well-thought-out planning to guarantee that the advantages are shared

Fig. 9.15 There is an extended vegetated buffer zone between the coastal areas and the urbanized areas in Karon

fairly. In the absence of this, adaptation may worsen socioeconomic inequalities and put the most vulnerable groups at higher risk.

Another lesson from Karon's experience is the importance of municipal cooperation in addressing persistent flooding by means of methodical infrastructure improvements. The significance of local political will and government continuity is demonstrated by the municipality's leadership in enforcing setback limits and implementing drainage improvements. Karon's shortcomings in replenishing natural buffers, however, also demonstrate that long-term resilience cannot be achieved only through manmade means. The takeaway from this is that even urban coastal areas with sound governance must broaden their adaptation plans to incorporate ecosystem-based solutions that may change in response to climate change and offer both ecological co-benefits and protection.

Another viewpoint is provided by Rawai's strategy, which is based on ecological stewardship and community-led action. Campaigns to raise awareness of hazards and restore mangroves are ingrained in local livelihoods here, particularly in the fishing industry. By combining adaptation with cultural and economic practices, resilience is produced that is not exclusively dependent on top-down governance or outside finance. However, Rawai's capacity to quickly implement extensive protective infrastructure in the face of immediate threats is restricted by its lack of financial and technical resources. In order to manage high-impact events, physical defence investments must be made in addition to social capital and local ownership, which

Fig. 9.16 A low-tide view from Rawai Beach, where exposed mudflats reveal the delicate balance between marine ecosystems and human activity. As coastal development intensifies, the challenge of managing erosion, tidal fluctuations, and ecological preservation grows more urgent, emphasizing the need for sustainable coastal management strategies in most of Phuket's beach areas

are essential for fostering adaptive capability. This is the lesson from Rawai (see Figs. 9.16, 9.17, 9.18, 9.19, 9.20, and 9.21).

The understanding—expressed or implied—that no one adaptation strategy is enough is a commonality among the three beaches. Hard infrastructure, nature-based solutions, land-use control, and public involvement are all combined in each location. However, there are differences in how these elements are balanced, which might affect resilience outcomes. Rawai places a higher priority on environmental restoration and community mobilization, Karon emphasizes improvements to municipal infrastructure, while Bangtao mostly relies on engineered constructions. The comparative lesson is that the best coastal management plans combine these components in ways that enhance one another, making sure that the advantages of one strategy balance out the disadvantages of another.

The distinctions also apply to stakeholder coordination and governance structures. Better coordination between resort-driven projects and state coastal management plans is necessary given Bangtao's reliance on private sector leadership. The municipal-led system in Karon demonstrates how public institutions can promote adaptation, but it runs the risk of becoming unduly preoccupied with short-term, observable infrastructure at the expense of the long-term health of the environment. Although local initiative fuels Rawai's grassroots strategy, it requires more robust

Fig. 9.17 An example of private land ownership and coastal development in Rawai beach

institutional backing to scale up and maintain adaption measures. This discrepancy highlights an important point: the governance framework that supports the measures implemented over time is just as important to the effectiveness of adaptation as the actual measures.

Lastly, the comparison analysis suggests a chance for cooperation and cross-learning among the three beaches on the island. Rawai's adaption of traditional customs can teach Bangtao how to do the same, which may increase community support for protective measures. Karon might benefit from Bangtao's experience fusing public priorities with private investment, which would enable it to diversify its sources of funding for adaptation. By modifying them to match its ecological and social settings, Rawai could gain from the technical know-how and engineering solutions tested in Karon and Bangtao. Phuket's coastal adaptation might be transformed from a collection of disparate initiatives into a cohesive, integrated resilience network through this kind of idea and practice exchange.

In sum, the parallels between Bangtao, Karon, and Rawai, serve as a reminder that they all confront risks that need for coordinated, systemic solutions. On the other hand, these distinctions offer a wealth of tactics and knowledge that can guide adaptive planning far beyond their own borders. The main takeaway is straightforward: resilience is the result of variety rather than homogeneity—of fusing different ideas, strengths, and methods into a cohesive, flexible whole that can handle the difficulties of a coastal environment that is changing quickly.

Fig. 9.18 Mudflats exposed at low tide in Rawai, where coastal residences sit directly at the edge of a fragile shoreline. The scene reflects the growing challenges of balancing waterfront development with the realities of tidal fluctuations, erosion, and ecological vulnerability in Phuket's coastal zones

Fig. 9.19 Here is the close interdependence between local livelihoods and the rhythms of the sea. These tidal shifts not only shape daily fishing practices but also highlight the delicate balance between community traditions and the pressures of changing coastal environments

9.5 Toward Integrated Coastal Resilience for Phuket's Future

Fundemanetally, the stories of Bangtao, Karon, and Rawai are stories of adaptation in motion—three different coastal communities that are connected by the same sea and influenced by the same climate forces yet manage resilience in ways that are unique to their respective economies, people, and geographies. The patterns are clear across these beaches: monsoon rains and seasonal flooding that put infrastructure to the test; storm surges that erode confidence and sand; rising seas that subtly reduce the capacity of existing defences; and human decisions, both past and present, that have changed the natural equilibrium between land and water.

Fig. 9.20 Here is an example of new development under construction with relatively shallow foundations. Private land ownership had led to extended private land development in costal areas of Rawai

Every facility has made progress toward increased security on its own. Although its reach is still unequal, Bangtao's resort-backed infrastructure has provided strong physical defences in high-value locations. Through organized drainage improvements and regulatory constraints, Karon's municipal cooperation has focused on persistent flooding. Rawai's community readiness initiatives and grassroots

Fig. 9.21 This scene struck me deeply: on one side, a local woman bends low, searching for sea worms to sustain her livelihood, while on the other, an international retiree (better known to be an international migrant) strolls leisurely along the coast with his partner and dog. In a single frame, Rawai's stark socio-economic contrasts are laid bare—between subsistence and leisure, survival and comfort—capturing the profound disparities that shape everyday life along many of Phuket's shores

mangrove restoration efforts have strengthened ecological and social resilience from the ground up. However, when combined, these strategies show that solitary intervention is insufficient to maintain Phuket's coastline's resilience. The responses must be interconnected with the threats.

Combining the advantages of all three situations is necessary for an integrated approach to coastal resilience. The importance of mobilizing private sector resources to finance top-notch infrastructure and upkeep is the lesson to be learned from Bangtao. Karon emphasizes the significance of uniform governance and the application of laws pertaining to risk-sensitive development. According to Rawai, resilience is a lived, shared duty rather than a foreign idea imposed from without. This is achieved by deeply ingraining adaptation into culture and livelihood practices. The island could create a cohesive model for coastal protection that is both strong and flexible if three advantages were combined: engineering and ecological restoration, corporate investment and public monitoring, and top-down planning and community-led initiative.

Such integration necessitates a governance structure that can coordinate across administrative borders and economic sectors, not merely well-meaning intentions. Harmonizing zoning regulations, coordinating hazard mapping, and establishing uniform guidelines for constructed and ecosystem-based solutions are all possible with a single coastal management plan for Phuket. It might make it easier to monitor coastal conditions jointly, allowing early warning systems to operate across municipal boundaries. Additionally, it might create mechanisms for resource sharing, guaranteeing that technical and financial assistance reaches communities whose resilience depends on the health of natural systems as well as the most commercially lucrative places.

Rethinking how harsh and soft measures are balanced will also be important for Phuket's shoreline in the future. Investments in mangrove belts, dune systems, coral restoration, and seagrass protection must be made in addition to seawalls and drainage systems, which are crucial for immediate hazard reduction. In addition to absorbing wave energy and lowering flooding, these natural resources support fisheries, boost biodiversity, and attract environmentally conscious tourists. They are living infrastructure; they develop, adapt, and, with proper maintenance, gain value over time. It is just as much of an economic and environmental investment to preserve and restore them.

The role of public interaction is equally significant. Poor preparation can weaken the strongest infrastructure, and in the absence of enabling infrastructure, even the best community mobilization can fail. For this reason, catastrophe awareness, evacuation planning, and seasonal readiness drills need to be ingrained in Phuket residents' daily lives. Understanding local dangers and involving locals, business owners, and tourists fosters a culture of resilience that cannot be accomplished by a single project or policy alone.

One conclusion emerges from the comparative analysis of Bangtao, Karon, and Rawai: resilience is a continuous process of adaptation, learning, and collaboration rather than a destination. Changes in the global economy, local development decisions, and climate change will all continue to influence the dangers. Whether Phuket will continue to address threats piecemeal or take advantage of the chance to develop a cohesive, integrated policy that protects its people and its natural heritage is the question.

The island may become more than a collection of isolated success stories if it can overcome its divisions between the public and private sectors, the urban and rural areas, and the engineered and ecological. It can serve as a model of coastal resilience for island economies around the world, demonstrating that, when coordinated, a variety of approaches can be an asset rather than a liability. The tide is still rising for Phuket's future, but so is the likelihood that we will be able to face it together.

References

Antoh, D. (2021). *Development strategy Phuview as leading tourist destination Bang Tao Phuket Thailand* (Doctoral dissertation, Sekolah Tinggi Pariwisa Ambarrukmo Yogyakarta).

Buakhao, S., & Weerakit, N. (2013). The impact of longstay tourists on local community: A comparative study between Tambon Rawai and Tambon Cherngtalay, Phuket, Thailand. *Review of Management Innovation & Creativity, 6*(19), 51–66.

Cheshmehzangi, A. (2025a). Innovation and environmental sustainability: ICT and beach management. In *Urban innovation and sustainability: ICT-based interventions in Hainan, China* (pp. 59–82). Springer Nature Singapore.

Cheshmehzangi, A. (2025b). The importance of health-sustainability-peace nexus in making Resilient Cities. *Peace and Sustainability, 1*(2), Article 100012.

Cheshmehzangi, A., He, B., & Zuo, T. (2025). Strengthening health-resilience nexus through sustainable urban design in Brisbane, Australia. In *Designing healthy cities: Integrating climate-resilient urbanism for sustainable living* (pp. 163–184). Springer Nature Singapore.

Choowong, M., Murakoshi, N., Hisada, K. I., Charusiri, P., Charoentitirat, T., Chutakositkanon, V., Jankaew, K., Kanjanapayont, P., & Phantuwongraj, S. (2008). 2004 Indian Ocean tsunami inflow and outflow at Phuket, Thailand. *Marine Geology, 248*(3–4), 179–192.

Kessaratikoon, P., Boonkrongcheep, R., Benjakul, S., & Youngchauy, U. (2013). Specific activities and radioactive contour maps of natural and anthropogenic radionuclides in beach sand samples (Patong, Kamala, Kata, Karon and Nai Yang) after tsunami disaster in Phuket province, Thailand. *Journal of Radioanalytical and Nuclear Chemistry, 297*(2), 247–255.

Kiriwongwattana, K., & Waiyasusri, K. (2024). Spatial evolution of smart cities for sustainable tourism: A case study of Phuket Province, Thailand. *Geojournal of Tourism and Geosites, 55*(3), 1312–1320.

Martin, S. A. (2010). The surfer-lifesavers of Phuket. *Thailand Surfrider, 3*, 40–42.

Martin, S. A. (2020). *Coastal currents in Phuket*. Thailand Surfrider, 3 pages report. https://www.stevenandrewmartin.com/wp-content/uploads/2020/04/Steven-A-Martin-2010-Coastal-Currents-in-Phuket.pdf

Neef, A., Attavanich, M., Kongpan, P., & Jongkraichak, M. (2018). Tsunami, tourism and threats to local livelihoods: The case of indigenous sea nomads in southern Thailand. In *The tourism–disaster–conflict nexus* (pp. 141–164). Emerald Publishing Limited.

Otto, M. (2016). *Action to catastrophe: A study on post-tsunami recovery of small businesses in Karon beach, Phuket*. https://www.diva-portal.org/smash/record.jsf?pid=diva2:895011

Pongprayoon, T. (2006). *Development of Phuket beaches for sustainable tourism: A case study of Patong, Kata, Karon* (Doctoral dissertation, Prince of Songkla University).

Wongbusarakum, S. (2002). *The Urak Lawoi and the complexity of sustainable Resource use: The political ecology of change in the Adang Archipelago*. University of Hawai'i at Manoa.

Wongwattanakit, C., Lohatepanont, M., & Pornchaiwiseskul, P. (2017). A socio-demographic study of destination management for coastal tourism in Phuket. *Journal of Environmental Management & Tourism, 8*(4 (20)), 737–750.

Chapter 10
Navigating Thailand's Tropical Thresholds: Toward Resilient and Integrated Coastal Futures

10.1 Understanding Thailand's Tropical Thresholds: Key Findings and Reflections

Insights from the previous case studies, thematic analysis, and conceptual debates are synthesized in this last chapter. It emphasizes how Thailand's coastal areas—especially those along the southern and eastern corridors—are at crucial junctures in terms of their institutional, social, and environmental conditions. Here, climate resilience is more than just a managerial or technical problem; it's a more profound chance for sustainable urban change (Cheshmehzangi, 2025) that is ingrained in regional ecosystems, customs, and governance methods. The chapter breaks down important lessons, guiding ideas, and useful takeaways for communities, scholars, planners, and policymakers navigating the unknown future via this lens.

In this section, the book's main idea of tropical thresholds is revisited, and the convergent pressures of climate change, fast urbanization, industrialization, ecological degradation, and socioeconomic disparity are discussed in relation to southern and eastern coastal Thailand. It makes the case that the region's distinct topography and social dynamics amplify its risks and resilience-building prospects. Thailand's coastal areas along its eastern and southern corridors are dynamic and complex thresholds where socioeconomic pressures and climatic vulnerabilities meet (Bennett et al., 2015; Cheewinsiriwat et al., 2024), posing special opportunities and challenges for sustainable urban transformation. This book has shed light on how various actors and governance structures are navigating these "tropical thresholds" in the pursuit of resilience through an integrative analysis of a variety of case studies, including the southern provinces of Krabi, Nakhon Si Thammarat, and Phetchaburi, the provinces of Chonburi, Rayong, and Chachoengsao, and the larger coastal area surrounding Bangkok.

The crucial role that place-specific geography plays in determining patterns of vulnerability and resilience is one notable discovery. With its vast mangrove ecosystems, karst structures, and estuarine habitats, southern Thailand has unique

A. Cheshmehzangi, *Thailand's Tropical Thresholds*, Urban Sustainability,
https://doi.org/10.1007/978-981-95-4975-7_10

ecological difficulties. Sea level rise, stronger storm surges, and changed rainfall patterns are examples of climate-driven events that have made coastal infrastructure, tourism, agriculture, and fisheries more vulnerable. However, these difficulties also spur creative ecosystem-based adaptive tactics. Effective management of ecological systems itself can support sustainable resilience techniques, as demonstrated by mangrove replanting projects, seagrass regeneration, and coral reef-friendly marine zoning (Bhatiyasevi, 1997; Janjamdara & Rasmeemasmuang, 2025), all of which are graphically represented by case studies in Krabi and Phetchaburi.

The EEC provinces, on the other hand, demonstrate how aspirations for economic growth interact with environmental and climate demands. National policy goals have sped up urbanization and industrialization, which has increased local vulnerabilities and complicated institutions. Industrial development, port expansions, and intensive infrastructure investment have often come at the cost of local ecological integrity and social cohesion. Rapid industrialization, for instance, has exacerbated climate-related vulnerabilities including coastal erosion and flooding in Chonburi and Rayong by causing glaring ecological degradation and pollution. At the same time, local governance capabilities and planning autonomy have been limited by institutional rigidities, which are exemplified by the EEC Act's centralized structure (Kim & Aroonpipat, 2016).

However, at the municipal level, these institutional restraints also encouraged innovative governing practices. Chachoengsao exemplifies the vital significance of local agency and government flexibility through its opposition and community-driven alternative planning responses. The province's initiatives to preserve agro-ecological traditions and preserve eco-agricultural identity in the face of larger industrial pressures highlight the revolutionary potential of local governance innovation (Chen, 2020). Similar to this, Laem Chabang Municipality's community-based resilience projects and adaptive zoning policies demonstrate how local leadership can ingeniously reduce national-local tensions and balance ecological sustainability with economic growth when aided by community involvement and academic collaborations.

The importance of participatory planning techniques and community-based adaptation (CBA) methods is another crucial result (Sekine et al., 2009). Local ecological knowledge, traditional governance systems, and inclusive participatory procedures are crucial for long-term resilience, as demonstrated by communities like Ban Nai Nang village in Krabi Province. Intergenerational knowledge transfer, risk dialogues, and resilience hubs are all beneficial strategies that strengthen local adaptive capacity, empower communities, and promote social cohesion. These instances show how deeply ingrained local cultural contexts in bottom-up adaptation tactics offer potent avenues for long-term change that are impossible to accomplish by external policy interventions alone.

Furthermore, the intricate urban environment of Bangkok and the nearby coastal regions serves as an example of yet another crucial aspect of tropical thresholds: the interdependent vulnerability between metropolitan centers and outlying coastal villages. Policymakers and urban planners have been compelled to reevaluate conventional urban planning paradigms due to the challenges of urbanization, flooding, subsidence, and urban heat islands. Promising paths toward more resilient urban futures are shown by initiatives in peri-urban Bangkok that combine urban agriculture, green infrastructure, and hybrid water management techniques. However, integrated governance that crosses institutional scales, sectoral differences, and urban–rural divides is necessary for successful change.

Social inequality constantly shows itself as a vulnerability amplifier in every situation examined (Albertson Fineman, 2017). The effects of climate change are frequently disproportionately felt by marginalized groups, including women, young people, and communities with low incomes. On the other hand, as several southern and eastern locales have shown, inclusive planning techniques can greatly improve adaptive capacity. In addition to enhancing local resilience, gender-inclusive decision-making procedures, youth-led environmental monitoring programs, and community-driven digital innovation also advance greater social fairness.

It is evident from synthesizing these varied experiences that Thailand's tropical thresholds are essential places of transformation rather than just risk areas. Stakeholders in Thailand's coastal regions can transcend these thresholds toward resilient, sustainable futures by embracing social inclusion, community empowerment, institutional flexibility, and ecological purity. Thus, the book's insights provide a thorough road map for scholars, decision-makers, and communities aiming to solve climate issues in Thailand and other coastal locations that are just as vulnerable worldwide.

10.1.1 Revisiting Vulnerability, Resilience, and Transformation Dynamics

Throughout this book, the concept of vulnerability has evolved beyond just being exposed to environmental risks to become a multifaceted state influenced by historical marginalization, institutional fragmentation, ecological degradation, and socioeconomic inequality (Cheshmehzangi & Zou, 2025; Taheri et al., 2025). With their natural diversity, economic importance, and swift changes, Thailand's coastal regions in the south and east represent a confluence of long-standing and growing risks. Every region, from the industrialized shorelines of Rayong to the low-lying, flood-prone plains of Phetchaburi to the mangrove-buffered settlements of Krabi, narrates a distinct tale of risk, resiliency, and reaction.

Mangroves, coral reefs, seagrass meadows, estuary wetlands, and sandy coastal margins are some of the tropical marine ecosystems that characterize these areas ecologically. From carbon sequestration and cultural identity to fisheries production and shoreline protection, these ecosystems offer vital benefits. However, as multiple chapters have demonstrated, they are increasingly being impacted by anthropogenic factors like dredging, land conversion, and chemical discharge, as well as climate-induced stressors including rising seas, saltwater intrusion, and coastal erosion. In addition to being ecological, this deterioration has a ripple effect on health, revenue production, and food security, especially for communities whose livelihoods depend on these ecosystems.

The region's institutional landscape is complicated, with overlapping jurisdictions, inconsistent enforcement, and incoherent planning. Local priorities and national development goals, particularly those centered on the Eastern Economic Corridor (EEC), have frequently clashed, resulting in governance gaps that increase the dangers associated with climate change. Laem Chabang's hybrid governance innovations, Rayong's overworked municipal systems, and Chachoengsao's opposition to top-down industrial planning all serve as examples of how institutional autonomy and capacity directly influence the paths of local resilience.

Thailand's coastal areas face a conundrum in terms of their economy. They support ports, tourist destinations, industrial complexes, and fisheries, making them growth engines. However, they are also delicate economic ecosystems that are susceptible to changes in the global market, resource depletion, and climate variability. While salt-farmers and aquaculture communities in Phetchaburi manage narrow margins in increasingly saline and uncertain settings, the pandemic years revealed the precarity of tourism-reliant economies like Krabi. From climate-smart zoning to floating agriculture, adaptive solutions have surfaced, but their scope and reach are still unequal and frequently constrained by institutional will, funding, or technological know-how.

The picture is equally diversified on a social level. Resilience can be significantly facilitated by social cohesion, intergenerational knowledge transfer, and inclusive leadership, as communities like Ban Nai Nang have shown. Others deal with fragmentation, young exodus, and a lack of confidence in the government. The way resilience is lived and performed on the ground is influenced by gendered labour divisions, the function of faith-based leadership, and young digital engagement tactics. These are not incidental; rather, they determine whether adaptation succeeds or fails.

The book's chapters have also revealed moments of transformative opportunity amid these interconnected aspects of vulnerability. In this way, the concept of tropical thresholds serves as a framework for systemic change as well as a metaphor for environmental tipping points. Thresholds can be both places of disruption and reorientation in the context of Thailand's coastal regions; in these situations, crisis creates opportunities for experimentation, coalition building, and the creation of new development narratives.

The seeds of change are found in the routine adaptation activities, whether it is through community-based risk discussions in flood-prone estuaries, adaptive marine zoning close to Si Chang Island, or youth-led monitoring in Krabi. These methods promote an embedded, relational, and iterative approach to resilience, challenging linear, top-down approaches. Therefore, transformation involves rearranging relationships—with location, with power, and with one another—rather than letting go of the past.

10.1.2 The Function and Symbolism of Water Towers in Rural and Township Thailand

The tall, cylindrical silhouette of a water tower is a recurring feature on the horizon of rural Thailand and its smaller townships. Sometimes, it seems they were left behind in urbanization processes and new development zones. And sometimes, they are just part of the new urban landscapes, reminding us of the challenges that cannot be left hidden. These structures are more than just functional reservoirs; they are a part of the physical and social fabric of communities, representing both practical necessity and symbolic presence. To comprehend their role is to appreciate the intersection of infrastructure, resilience, and cultural meaning in Thailand's smaller settlements.

Water towers provide an essential technical purpose at the most basic level. Even in places that are not entirely served by sophisticated piped networks, they guarantee that populations have access to a consistent supply of clean water by acting as storage facilities. Water delivery is still erratic in many rural and peri-urban areas of Thailand because of inadequate infrastructure, seasonal variations in rainfall, or limited energy supplies. The towers' raised design lessens reliance on electric pumping systems by enabling water to flow into residences and community taps with adequate pressure. They provide efficiency and dependability by using the straightforward theory of gravity, especially in situations when power is expensive or erratic.

Water towers are closely related to resilience during times of crisis or scarcity, even beyond everyday convenience. During the dry season, there are frequently severe water shortages in Thailand's rural areas, especially in the Northeast. On the other hand, ground-level water systems may become overloaded or contaminated by monsoon floods. Both extremes are lessened by elevated towers, which protect uncontaminated supplies during floods and store reserves during droughts. In this way, they serve as vital buffers in a climate that is getting increasingly variable and harsh. Their presence anchors community resilience by directly supporting agricultural practices, public services like schools and health centers, and household well-being.

However, their use goes well beyond engineering and hydraulics. In Thailand, water towers have come to represent communal progress. They are frequently built using a mix of community work, local management, and government funding. Even in the tiniest settlements, they serve as tangible symbols of municipal advancement, or civic progress, and state presence due to their prominence on the skyline. A water tower can be seen as a pledge to deliver necessary services, protect livelihoods, and keep life going in the face of environmental uncertainty.

Interestingly, I somehow found out that these structures also serve as points of identity and orientation. I asked a few people and that's what some people said, too. Water towers can serve as landmarks, gathering places, or navigational aids in rural areas where other vertical construction methods are uncommon. Their symbolic significance is increased by their visibility across coastal lowlands, orchards, and rice fields. Water towers' function as nodes of local infrastructure is further enhanced when they are converted or shared with communication technology, housing antennae or signal devices.

However, water towers also raise questions about infrastructure vulnerability and future adaptation. Although they have historically offered resilience, in a time of changing rainfall patterns and aquifer stress, their reliance on adequate beginning water sources continues to be problematic. Furthermore, rural towers run the risk of being overlooked or receiving subpar maintenance as Thailand's water systems continue to modernize and become more urbanized. In many townships, the presence of rusting or abandoned towers is indicative of both uneven development trajectories and infrastructure shifts. Such imagery is poignant: once symbols of modern progress, some towers are now reminders of neglected rural investment or shifting priorities in resource allocation.

However, the larger story of water towers is one of change rather than loss. These structures might become even more crucial if climate change worsens and water security becomes a more urgent concern. Their ability to distribute, lift, and store water is in line with the expanding demand for flexible and decentralized infrastructure solutions. The dual nature of Thailand's rural resilience is thus embodied by water towers, which are both useful devices and timeless cultural icons.

In Thailand's rural and township areas (as well as in new urbanised zones and new development areas), the water tower symbolizes the meeting point of symbolic and function. It is a resilient artifact that was created out of need but has layers of significance that go beyond politics around infrastructure provision to social identity and community pride. These towers remind us that adaptation is frequently embedded in commonplace yet profoundly meaningful buildings by raising not only the visibility of water but also the visibility of local resilience. By going over them again, we acknowledge the modest yet vital role they still play in managing vulnerability, maintaining resilience, and influencing the processes of transformation in Thailand's varied landscapes (see Figs. 10.1, 10.2, 10.3, and 10.4).

10.2 Integrated Governance for Sustainable Urban Transformation

It takes a rethinking of governance itself to achieve sustainable urban development in Thailand's coastal regions, not only the implementation of technical solutions or climate finance systems. A recurring issue in all the case studies in this book is the necessity of integrated, flexible governance structures to close the gap between local reality and national aspirations. Governance structures need to be able to distribute power, manage complexity, and foster community-led innovation, especially in situations where ecological fragility and socioeconomic volatility coexist.

Tensions in governance are exemplified by the Eastern Economic Corridor (EEC). A centralized development vision is superimposed on various local landscapes by

Fig. 10.1 A water tower is now part of the urban landscape of outskirts of Chumphon in eastern Thailand

the EEC, a state-driven economic engine. The EEC Act gives central authorities a great deal of authority, frequently at the expense of local and provincial governments. Planning disjunctions have resulted from this, especially in environmentally sensitive areas where local development plans and national spatial goals clash. Local politicians and community organizations in Chachoengsao are fighting against industrial expansion into protected agricultural and ecological regions because of these tensions. The ramifications of not aligning vertical policy frameworks and the possibility of local governance actors reclaiming agency in defence of ecological and territorial identity are both illustrated by this scenario.

On the other hand, Ban Nai Nang provides a grounded, bottom-up adaptive governance approach. Here, a participatory planning culture focused on ecological stewardship has been developed by local leadership, religious organizations, and intergenerational knowledge networks. Youth-led monitoring, ecotourism, and mangrove restoration are locally designed solutions that are ingrained in the community's identity rather than being schemes that are imposed from without. This accomplishment is made possible by strong institutional relationships with universities, NGOs, and specific governmental institutions that uphold local sovereignty in addition to community willpower. This hybrid governance style ensures legitimacy and long-term sustainability by striking a balance between grassroots initiative and state backing.

Fig. 10.2 A water tower in new coastal development zones. This ensures a steady supply of water for households, schools, hospitals, and small businesses, especially where piped water networks are limited or unreliable

In this setting, integrated governance entails more than just coordination; it calls for multi-scalar, inclusive, and reflexive systems. As demonstrated by Laem Chabang's attempts to include climate resilience into zoning plans through partnerships with academic institutions, institutional collaboration must be combined with policy consistency. Formalizing multi-level accountability is also necessary to guarantee that climate hazards are addressed at all administrative levels and that local actors and central authorities respond to one another.

Finally, the Thai experience shows that institutional learning and context sensitivity, rather than governance consistency, are what determine sustained urban development. Governance becomes a facilitator of resilience rather than a barrier when national frameworks recognize local specificity, when local perspectives influence policy creation, and when intermediary actors cultivate trust at all levels.

Fig. 10.3 A water tower is functional that helps maintain a safe and uncontaminated water supply, yet a landmark in townships and villages visible from far away,

Fig. 10.4 During floods or storms, when ground-level water systems may be contaminated or damaged, the water towers are sometimes linked with community disaster resilience planning

10.3 Place-Based and Participatory Adaptation as a Foundation for Long-Term Resilience

This section makes a compelling case for the importance of locally based adaptation solutions by heavily referencing case studies, particularly the Ban Nai Nang example in Krabi Province. It highlights that participatory planning, inclusive community practices, place-based ecological knowledge, and intergenerational knowledge transfer are essential components of effective coastal climate resilience. It offers a convincing summary of how community-driven resilience can become the cornerstone of sustainable transformation strategies by reflecting on creative practices highlighted throughout the chapters, such as ecosystem-based adaptation, youth-led digital monitoring, gender-inclusive decision-making, and religiously informed stewardship.

Thailand's coastal areas cannot be made resilient by national-level planning or by relying only on infrastructure. Sustainable climate adaptation must come from within communities, woven into local ecosystems, social fabric, and lived experience, as this book's chapters demonstrate. Participatory and place-based adaptation is not merely a technique; it is essential. The most reliable and long-lasting route to long-term resilience is adaptation that is based on local knowledge and co-produced through inclusive involvement, particularly in areas characterized by ecological sensitivity, cultural specificity, and institutional diversity.

Krabi Province's Ban Nai Nang provides a strong foundation for this claim. Amidst karst rocks and mangrove estuaries, this rural town with a majority of Muslims has created a climate resilience strategy that is firmly anchored in the area. How the community interprets, responds to, and foresees environmental change is shaped by the generational transmission and preservation of local ecological knowledge about seasonal fisheries, mangrove dynamics, and tidal shifts. In this case, adaptation is not abstract. Restored mangrove belts that guard against storm surges, ecotourism projects that sustain livelihoods while preserving biodiversity, and family-based skill transfer that guarantees ecological stewardship is practiced for generations to come are some examples of how it is manifested.

Ban Nai Nang's participatory planning procedures have been similarly fundamental. A common awareness of local vulnerabilities and priorities has been established through community-led resilience centers, inclusive risk conversations, and village mapping projects. Crucially, these practices have given women, young people, and the elderly—groups who are frequently left out—the ability to influence decisions. The values of respect for one another and local authority have served as the foundation for the community's interactions with outside parties including NGOs, researchers, and representatives of the local government. Not only for the community, but by it, this is adaptability.

This philosophy is further supported by other chapters. Communities that grow salt have combined biodiversity-friendly farming methods with mangrove restoration in Phetchaburi. Among Rayong, youth-led projects monitor the water and air quality in industrial areas using smartphone apps, fostering awareness and a sense of agency among the next generation. Opposition to the EEC's top-down industrialization in Chachoengsao has sparked alternative planning narratives that protect rural livelihoods and restore ecological uniqueness. One thing is evident in all these diverse settings: adaptation succeeds when it fits in with the rhythms, values, and aspirations of place.

It has become clear that intergenerational knowledge plays an especially important role. In Krabi and Phetchaburi, elders serve as living repositories of ecological knowledge, frequently having managed fishing grounds and tidal cycles for decades. Their wisdom is dynamic, though. Through narratives, religious instruction, educational initiatives, and online forums, it is disseminated, discussed, and reinterpreted. In turn, young people have emerged as innovators, utilizing participatory media to record local climate tales, deploying drones to track mangrove recovery, or starting online campaigns for plastic-free coastlines. Through intergenerational interaction, adaptive ability is developed that is resilient to changing circumstances, culturally significant, and legitimate locally.

Crucially, the participatory tactics discussed in this book are neither symbolic nor isolated. They stand for intentional design decisions—methods to operationalize agency distribution, operationalize inclusion, and integrate resilience into preexisting social structures. For instance, gender-sensitive methods were not optional. The acknowledgement of women's contributions to informal education, caregiving, and household economies influenced community responses to climate change in areas like Ban Nai Nang. Likewise, mosque-based programs and religious leaders provided social cohesiveness and moral authority that supported adaptive behaviour.

Conventional planning paradigms that depend on strict cost–benefit evaluations, technocratic forecasting, or the replication of generic models are challenged by these cases. They contend that adaptation needs to be iterative, culturally based, and co-created. The implication is significant. Scaling sameness is not the similar as scaling up adaptation. It entails providing a range of context-specific paths, reinforced by accommodating financial sources and supportive policy frameworks.

National frameworks must change from imposing blueprints to enabling scaffolds in order to accomplish this. In addition to infrastructure, ministries and development organizations must invest in processes, such as communication, facilitation, capacity building, and trust-building. Academic institutions and non-governmental organizations need to acknowledge their positions as co-learners and facilitators of indigenous wisdom, rather than as experts imparting knowledge. Instead of just implementing, municipalities need to be encouraged to get together and respond.

Place-based and participatory adaptation is fundamentally a power shift. Who decides, who knows, and who gains are the questions it poses. Thailand's coastal villages serve as a reminder that true resilience is constructed from the inside out, from mangrove roots to human networks, in the face of rising seas, erratic storms, and unequal development demands. The future of sustainable transformation really resides in these inclusive, co-owned, and grounded methods.

10.4 Strategic Directions and Action Checklist for Thailand's Coastal Climate Resilience

The need to rethink climate resilience in coastal Thailand as a locally based, socially inclusive, and ecologically grounded transformation unites the book's various instances and topic studies. This journey through Thailand's tropical thresholds has produced a rich collection of strategic approaches, ranging from industrializing shorelines in the Eastern Economic Corridor to adapting to rural villages in the southern provinces. Researchers, educators, civil society organizations, and community leaders who work together to design the resilience pathways will also find value in these lessons, in addition to planners and policymakers (see Fig. 10.5).

The importance of ecosystem-based adaptation (EbA) as a pillar of climate resilience (Tiwari et al., 2022) is one of the most obvious themes across the chapters. In addition to being natural resources, coastal ecosystems like mangroves, coral reefs, and seagrass beds serve as living infrastructure that protects against storm surges, prevents erosion, sustains fisheries, and stores carbon. Restoring and co-managing these ecosystems can produce resilience dividends that go beyond environmental advantages, bolstering livelihoods, social cohesion, and place-based identity, as demonstrated by initiatives in Krabi, Phetchaburi, and the island settlements of Si Chang. Therefore, EbA investment needs to go beyond pilot projects and become a key component of Thailand's growth goal and national adaption strategy.

The institutionalization of integrated governance approaches is equally important. Coherent climate action has frequently been hampered by the division between national economic priorities and subnational environmental obligations, as is exemplified by the EEC's complex governance. Laem Chabang's innovative municipal interventions and the local resistance to industrial encroachment in Chachoengsao

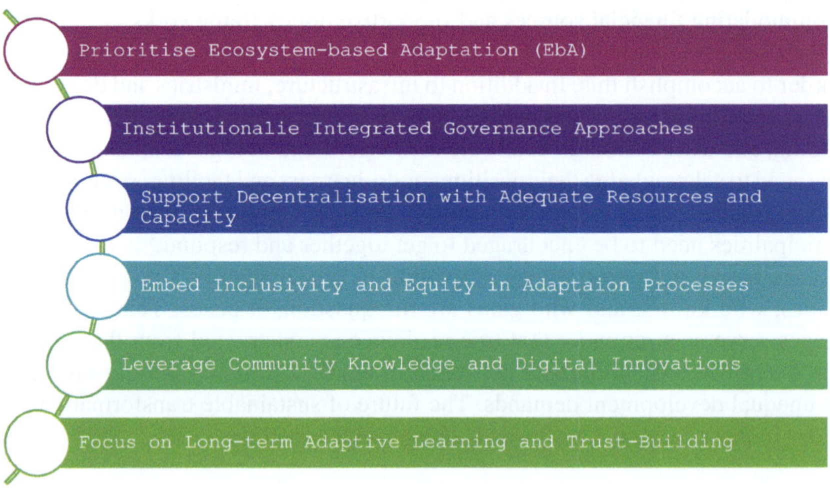

Fig. 10.5 Summary of checklist of 6 key messages and takeaway ideas from the book

demonstrate the prospects and difficulties of bringing various levels of governance into alignment. Mechanisms that link provincial planning, community-based initiatives, and national strategies are necessary for integrated governance. In order to guarantee that resilience planning is not isolated but rather integrated throughout institutional landscapes, it necessitates policy consistency, vertical coordination, and cross-sectoral collaboration.

In fact, decentralization—a stated goal in Thailand's policy architecture—remains inconsistent. Limited technical skills, tight finances, and unclear planning mandates are problems for secondary cities and municipalities, especially those in ecologically vulnerable or fast urbanizing areas. Municipalities in Rayong, for example, face severe capacity shortages despite being disproportionately affected by climate-related hazards. The promise of decentralized adaptation won't be fulfilled unless local authorities are given the tools, instruction, and clear legislation they need. Thus, concrete support systems that allow local governments to take the lead on climate resilience and are adapted to their institutional, cultural, and geographic settings must be implemented alongside national reforms.

Additionally, adaptation needs to be viewed as a process that is essentially inclusive. Cultural variety, young autonomy, and gendered knowledge systems are not incidental factors; rather, they are vital sources of adaptive potential. The book has shown how social equity improves ecological and institutional resilience, from the importance of women's responsibilities in care-based resilience in Ban Nai Nang to the youth-led digital monitoring of environmental change in Rayong. Frameworks for climate planning must proactively take these viewpoints into account when creating participatory methods that go beyond consultation to include shared leadership and decision-making. Inclusion is a practical requirement for fair and efficient adaptation as well as a normative objective.

Additionally, the combination of contemporary digital innovation and traditional ecological knowledge has become a revolutionary force that transcends geographical boundaries. The rich, generation-old knowledge that communities have about seasonal fisheries, monsoon cycles, and tidal rhythms can and should guide the creation of scientific models and policies. Younger generations are now able to participate in local planning, lobbying, and real-time monitoring because to the development of digital tools including smartphone apps, participatory mapping platforms, and online storytelling. A viable path toward improving climate governance's responsiveness, transparency, and democratic participation is represented by these hybrid knowledge systems.

Last but not least, the most crucial strategic takeaway from this book is the necessity of viewing climate adaptation as an ongoing, iterative process based on mutual respect and learning. Whether in the agro-ecological planning of Chachoengsao or the peri-urban flood management zones of Bangkok, it is evident that resilience cannot be forced or finished in a single project. Rather, it needs to be fostered via consistent involvement, testing, and introspection. This calls for adaptable financing plans, educational institutions, and a caring mindset—care for ecosystems, knowledge passed down through generations, and the social relationships that provide communities resilience. Establishing trust among various stakeholders, between the state and society, and across sectors is not a side project; rather, it is the foundation of adaptation.

All things considered, the lessons learned from Thailand's coastal regions point to a compelling conclusion: resilience is a social-ecological process rather than a technical solution. Mangrove roots, village meetings, contentious planning rooms, and the imaginations of young people with smartphones and elderly people with memories are all part of its construction. Thailand can not only confront but also transform through the thresholds it faces by focusing on place, empowering communities, reforming governance, and investing in education.

10.5 Final Reflections: Navigating Forward Through Uncertain Thresholds

Reflecting on the idea of "tropical thresholds" as dynamic regions of change, this final section brings the volume to a close. It restates the main point, which is that coastal resilience in Thailand rests on a thorough comprehension of these thresholds and treating them as chances for revolutionary change rather than just risks being reduced. In order to respect the diversity and complexity of coastal communities, ecosystems, and urbanizing regions, the section emphasizes the urgent need for nuanced, context-sensitive approaches.

Finally, Thailand's tropical thresholds are invitations rather than obstacles. They challenge stakeholders, communities, legislators, and planners to reconsider how they relate to governance systems, ecosystems, and one another. They encourage innovation in relationships, trust, and teamwork in addition to technologies and infrastructures. They encourage a shift from limited ideas of resilience as survival to resilience as thriving, which entails building communities that can deal with uncertainty in a creative, cooperative, and compassionate manner.

The southern and eastern coastal regions of Thailand may pave the way for truly sustainable urban futures by adopting the integrated governance, place-based knowledge, participatory procedures (Cheshmehzangi & Dawodu, 2018), inclusivity, and continuous learning principles discussed throughout this book. In addition to diagnosing vulnerabilities, this book provides a blueprint of possibilities that are rooted in local reality yet have global resonance. The information presented here provides vital direction for negotiating thresholds into resilient, inclusive, and regenerative futures as coastal Thailand continues to face environmental, social, and economic change (see Figs. 10.6, 10.7, 10.8, 10.9, and 10.10).

Going forward, Thailand's coastal resilience will depend more on having the guts and ability to deal with the unpredictable than it will on being an expert at the predictable. By their very nature, tropical thresholds are resistant to easy fixes; instead, they call for innovative thinking, intergenerational cooperation (Su et al., 2024), and flexible thinking. A new ethic of coexistence will be necessary for the future, one in which infrastructure is built with empathy in addition to efficiency,

Fig. 10.6 I stood here near Cha-Am in the sweltering midday heat, contemplating the quiet yet inevitable signs of decay—an unsettling reminder of how fragile both land and community can be when good governance falters. The scene evoked not only the weight of neglect but also the looming possibility of disappearing villages, eroded coastlines, and lost opportunities. In that moment, I found myself hoping for a different future: one where these places are not reduced to scattered and patchy islands along the southeast coast, but instead emerge as vibrant local communities, grounded in resilience and enriched by prosperity and joy. The challenge lies not in imagining such a future, but in building it—through care, foresight, and the collective will to ensure that these coastal settlements endure, adapt, and thrive against the pressures of climate change, urban expansion, and governance shortfalls

policy is formed by discussion rather than data, and resilience is gauged by renewal rather than recovery. The true revolution will come from how we choose to react—not in fear, but with creativity, humility, and a common goal—as the tidelines change and the climate puts our systems to the test. Let these boundaries mark the start of something bigger: a reinterpretation of development as determined by our capacity to live sensibly, fairly, and peacefully within the cycles of coastal life.

Fig. 10.7 Along this quiet riverine stretch, where mangroves and nipa palms meet the edges of human settlement, the delicate interplay between natural ecosystems and human interventions becomes visible. The calm waters mask the underlying tensions—seasonal floods, creeping salinity, and the encroachment of development—that shape Thailand's tropical thresholds. Rock embankments signal attempts at control and stabilization, yet the resilience of these landscapes lies equally in the living roots of mangroves that buffer the shoreline, filter the waters, and sustain biodiversity. Standing here, one is reminded of both vulnerability and promise: the risk of losing ecological balance if governance neglects integration, and the opportunity to forge coastal futures that align human livelihood with ecological continuity. This place calls for a vision of resilience not as a defensive wall but as a living partnership with nature, essential for thriving communities along Thailand's coasts

Fig. 10.8 A riverine edge near coastal settlements where mangroves and human-built embankments meet, reflecting both the vulnerability and resilience of Thailand's tropical thresholds

Fig. 10.9 This waterway captures the complex dynamics at play along Thailand's coastal thresholds, where natural systems and human interventions continuously negotiate space and survival. The stillness of the water belies its volatility, shaped by seasonal monsoons, saline intrusion, and shifting flows that test both ecological and social resilience. Such places remind us that the future of Thailand's coasts depends not only on hard infrastructure but on maintaining and integrating the ecological foundations that sustain life. Here, the coexistence of human and natural systems reveals both their vulnerability and their potential synergy in shaping resilient coastal futures

Fig. 10.10 Coastal waterways framed by mangroves and settlements, highlighting the fragile balance between ecological protection and expanding human activity along Thailand's tropical coasts

References

Albertson Fineman, M. (2017). Vulnerability and inevitable inequality. *Oslo Law Review, 4*(3), 133–149.

Bennett, N. J., Dearden, P., & Peredo, A. M. (2015). Vulnerability to multiple stressors in coastal communities: A study of the Andaman Coast of Thailand. *Climate and Development, 7*(2), 124–141.

Bhatiyasevi, U. (1997). Marine resources and fisheries in the exclusive economic zones of Thailand. In *Fishery resources and state of stocks exploitation in the Waters of the Gulf of Thailand, East Coast of Peninsular Malaysia and Andaman Sea* (pp. 1–58). Training Department, Southeast Asian Fisheries Development Center.

Cheewinsiriwat, P., Langkulsen, U., Lertwattanamongkol, V., Poompongthai, W., Lambonmung, A., Chamchan, C., Boonmanunt, S., Nakhapakorn, K., & Moses, C. (2024). Assessing coastal vulnerability to climate change: A case study of Nakhon Si Thammarat and Krabi. *Social Sciences, 13*(3), 142.

Chen, W. (2020). The transformation of agricultural production system in line with "Resource and Ecologically Sound Agriculture". In *Challenges and opportunities for Chinese agriculture: Feeding many while protecting the environment* (pp. 183–340). Springer Singapore.

Cheshmehzangi, A. (2025). The importance of health-sustainability-peace nexus in making resilient cities. *Peace and Sustainability, 1*(2), Article 100012.

Cheshmehzangi, A., & Dawodu, A. (2018). Stakeholder methods to climate change and sustainable development. In *Sustainable urban development in the age of climate change: People: The cure or curse* (pp. 23–68). Palgrave Macmillan.

Cheshmehzangi, A., & Zou, T. (2025). The importance of regional food security in achieving supply chain resilience: A case study. In *Resilient urbanism* (pp. 115–128). Routledge.

Janjamdara, A., & Rasmeemasmuang, T. (2025). Vulnerability of the upper coast of the Gulf of Thailand. *Engineering Journal, 29*(5), 25–35.

Kim, H., & Aroonpipat, S. (2016). *AEC preparation under recentralized Thai government: Case study of Chonburi and Rayong* (Doctoral dissertation, Thammasat University).

Sekine, H., Fukuhara, K., Uraguchi, A., Tan, C. K., Nagai, M., & Okada, Y. (2009). *The effectiveness of community-based adaptation (CBA) to climate change.* GEIC Working Paper Series 2009-001.

Su, Z., Bentley, B. L., McDonnell, D., Alimu, T., Dai, W., Jiang, J., Liu, Y., Yu, X., Feng, Y., Diao, Y., & Wu, X. (2024). Senior employment technology: Leveraging opportunities in population ageing. *Health Policy and Technology, 13*(3), Article 100883.

Taheri, S., Shabani, A., Badland, H., Sharifi, A., & Cheshmehzangi, A. (2025). Protecting our cities against future pandemics: A systematic review for creating health-supporting and resilient cities. SSRN 5055867.

Tiwari, A., Rodrigues, L. C., Lucy, F. E., & Gharbia, S. (2022). Building climate resilience in coastal city living labs using ecosystem-based adaptation: A systematic review. *Sustainability, 14*(17), 10863.

Chapter 11
From Threshold to Catalyst: Thailand's Role in Regional and Global Climate Futures

11.1 Understanding Regional and Global Climate Futures in Coastal Zones

The story of Thailand's tropical thresholds is far more than a national narrative. Fundamentally, it examines how communities located on the front lines of climate change are adjusting, improvising, and changing in response to global shifts. The world's coastal regions are the most obvious—and significant—places for this conflict to occur. These transitional areas, where land and sea meet and natural processes and human activity combine, are some of the planet's most productive and most endangered ecosystems. Large populations, vital infrastructure, and regional economies are all supported by them, but they are also the first to experience the destabilizing effects of storms, increasing sea levels, saltwater intrusion, and ecosystem degradation.

Because it exemplifies the intricate interaction of climate, geology, society, and economics that characterizes so many coastal places worldwide, Thailand's coastal experience provides an especially insightful view into the future. With its extensive networks of fisheries, tourism, and urban growth corridors, densely populated delta regions, and long shorelines along the Andaman Sea and the Gulf of Thailand, the nation faces challenges similar to those of its neighbours, including Myanmar, Cambodia, Vietnam, and Malaysia, as well as those of more distant countries, including Bangladesh, Indonesia, and the Philippines (See the next sub-section for further discussion). Therefore, it is not just of national interest to comprehend how Thailand manages its coastal thresholds; it is also essential to comprehending the more general trends of climate resilience in the Global South and beyond.

The scale of the stakes cannot be overstated. Nearly 40% of the world's population lives along the coast, and many of the cities with the greatest rates of growth are found around low-lying shorelines. Despite being centers of biodiversity, trade, and cultural interaction, these areas are becoming increasingly vulnerable to climate-related threats. Under current emissions trajectories, sea levels are predicted to rise

A. Cheshmehzangi, *Thailand's Tropical Thresholds*, Urban Sustainability,
https://doi.org/10.1007/978-981-95-4975-7_11

by up to one meter by the end of the century, endangering the habitability of entire regions. At the same time, storm surges are moving farther inland, coastal erosion is speeding up, and tropical storms are becoming more frequent and stronger. In addition to altering physical landscapes, these forces are upending socioeconomic structures and compelling communities, governments, and businesses to reconsider their long-term plans.

Thailand's coastal adaptation story speaks directly to these global dynamics. The establishment of community-based disaster preparedness programs in fishing villages, the construction of mangrove buffer zones in provinces like Krabi and Nakhon Si Thammarat, and the renovation of urban coastal infrastructure in cities like Bangkok and Pattaya are all examples of how a country can use ecological knowledge, social capital, and institutional innovation to adapt to and even benefit from change. These interventions show how reactive strategies, like erecting seawalls or uprooting communities, are giving way to more proactive, systemic strategies that incorporate participatory governance, nature-based solutions, and adaptive design concepts. These tactics strike a deep chord with related projects that are beginning to emerge in the delta regions of Bangladesh, the Mekong Delta of Vietnam, and the island communities of Indonesia, demonstrating a common set of resources and methods that cut across national borders.

Yet the value of Thailand's example extends beyond practical measures. It also offers conceptual lessons about how we frame and respond to climate futures in coastal contexts. The discourse surrounding the global climate too frequently views vulnerability as a passive state that needs to be avoided. The strategy used in Thailand reinterprets vulnerability as a dynamic area of experimentation where new kinds of cooperation, creativity, and resilience-building are sparked by the demands of climate change. For other areas under comparable constraints, this reinterpretation has significant ramifications. For instance, methods modelled after Thailand's participatory adaptation models could guide more inclusive and locally based resilience tactics in the Bay of Bengal, where populations are frequently uprooted by cyclones and sea level rise. Thailand's integrated coastal planning and ecosystem restoration initiatives could serve as a model for tiny island states facing existential challenges in the Pacific as they work to protect their natural and cultural heritage.

Furthermore, by connecting local efforts to regional and global frameworks, the Thai experience emphasizes the value of cross-scale thinking. Resilience cannot be developed in a vacuum, because climate change transcends national boundaries. Coastal adaptation necessitates a multi-tiered governance structure that links village-level projects to international climate funding networks, regional cooperation mechanisms, and national policies. Thailand's involvement in international adaptation funds, its involvement in ASEAN environmental frameworks, and its partnerships in research and development are all components of a larger ecosystem of action that demonstrates how regional responses may support global climate agendas. To handle the cascading hazards and systemic changes of the upcoming decades, this integrated approach is exactly what is required.

In many ways and not just limited to the scope of this book, Thailand is essentially recognized as a living laboratory where solutions are being tested, improved, and scaled, a microcosm of the larger coastal climate dilemma. Therefore, the lessons learned from its tropical thresholds are not limited to the boundaries of the country. They highlight the common risks and advantages faced by coastal communities throughout Asia, Africa, and the Pacific. More significantly, they offer a lexicon of resilience based on co-evolution, diversity, and adaptability, elements that will be crucial as humanity negotiates the choppy waters of the Anthropocene.

As we look toward the future, the question of whether coastal zones will change in the future has been replaced with the question of how societies will influence that change. Thailand's experience indicates that the best solutions will be those that welcome change as a chance to create more resilient, fair, and environmentally conscious futures rather than trying to prevent it. In this way, the teachings coming from its coasts are both incredibly local and incredibly global; a reminder that while the road to planetary resilience may start in the specific, it must always extend beyond.

11.1.1 Beyond Thailand: Motivations, Relevance, and Global Resonance

This book was inspired by a pressing need to investigate how communities in the tropics, which are at the forefront of climate change, might adapt to significant changes in the environment and society. The coastal areas of Thailand offer a distinctive and illuminating backdrop for this investigation. They embody the challenges faced by many low- and middle-income nations that are quickly urbanizing, industrializing unevenly, and coping with climate changes that surpass institutional capability; they are neither severe outliers nor isolated exceptions. Therefore, Thailand is more than just a location; it serves as a prism through which more general trends of susceptibility, adaptability, and change can be seen on a global scale.

A number of considerations led to the choice to concentrate on Thailand. First, it is a crucial test site for comprehending resilience in tropical coastal ecosystems due to its topography and climate. Thailand exemplifies the challenges of managing coastal change in the Global South with its more than 3000 km of coastline, diverse ecosystems that include mangroves, coral reefs, deltaic plains, estuaries, and quickly expanding coastal towns and cities. Second, the country's developmental trajectory is similar to that of many Southeast Asian countries, with a middle-income economy striking a balance between environmental fragility and economic aspirations. Examining how climate adaptation might be economically feasible, ecologically sound, and socially inclusive is made possible by this dichotomy of vulnerability and aspiration.

The lessons drawn from Thailand resonate far beyond its borders. Countries in South and Southeast Asia, including Bangladesh, Cambodia, Myanmar, and Malaysia, face remarkably comparable issues. Mangrove deterioration, rising seas, and land subsidence are colliding with Malaysia's fast coastal urbanization around Penang and Johor Bahru, reiterating the conflicts seen in Chonburi and Rayong. Similar circumstances to those in Phetchaburi and Nakhon Si Thammarat are being created in Myanmar's Ayeyarwady Delta by increased cyclone danger, erosion, and salinization. Similar to how southern provinces in Thailand must balance ecological preservation with economic necessity, the Tonle Sap basin and coastal provinces of Cambodia are grappling with how to balance hydrological instability with urban and agricultural demands. Perhaps the most striking analogy is offered by Bangladesh, whose deltaic coastlines, which are vulnerable to mass displacement, storm surges, and sea level rise, reflect many of the vulnerabilities seen in Thailand's low-lying coastal zones. However, they also show how large-scale community-based adaptation and nature-based solutions can be scaled nationally.

These parallels are not coincidental. They result from institutional equivalents and similarities, such as colonial legacies that influenced governance and land use, development plans that put growth ahead of ecological balance, and a shared susceptibility to sea level rise, tropical cyclones, and monsoon variability. They also mirror more fundamental systemic problems, such as insufficient adaptive capacity, resource inequity, and institutional fragmentation, all of which influence the course and speed of resilience-building initiatives. To demonstrate how case-specific insights might guide adaptation techniques elsewhere, the book places Thailand's experience within this larger regional context.

Thailand's ecosystem-based adaptation experiments, such as the restoration of mangroves in Krabi and the creation of coral-friendly marine zoning in Phetchaburi, provide scalable models for managing coastal ecosystems that may be advantageous for deltaic Bangladesh or the mangrove-rich southern Myanmar coasts. Lessons in community-led planning can be directly applied to Cambodian provinces that are struggling against extractive development, as demonstrated by the governance innovations in Chachoengsao, where local authorities developed agro-ecological alternatives and opposed top-down industrial expansion. In a similar vein, Thailand's focus on participatory adaptation, which is evident in Ban Nai Nang's incorporation of intergenerational stewardship, religious leadership, and local knowledge, demonstrates how cultural specificity can foster resilience. This is a lesson that multicultural societies like Malaysia and Indonesia can learn from.

Beyond Southeast Asia, the Thai experience can be applied to coastal cultures in the tropics and subtropics, from West African coastal cities struggling with erosion and displacement to Pacific Island states facing existential climatic risks. The central idea of this book, the concept of tropical thresholds, is a globally applicable framework rather than a localized metaphor. It highlights the intersections of development pressure, social vulnerability, and ecological fragility, which can result in drastically different transformational paths. Whether in the low-lying settlements of Fiji, the flood-prone deltas of Mozambique, or the rapidly urbanizing mangrove coasts of Brazil, Thailand's lessons about participatory planning, ecological stewardship, flexible governance, and the blending of traditional and scientific knowledge are still highly relevant.

Crucially, the predominance of Global North viewpoints and solutions in climate adaptation research is another prevalent bias that this book aims to address. All too frequently, technologically complex, expensive initiatives that are impractical and inappropriate in many regions of the world are used to frame resilience. This book adds to a growing body of research that repositions tropical and developing environments as areas of innovation rather than just vulnerability by focusing on Thailand, a nation that is firmly in the Global South but has amazing variation in adaptation behaviour. Communities in Thailand show that effective adaptation frequently results from collective action, cultural continuity, and ecological reciprocity; principles that are universally applicable, rather than from sophisticated infrastructure or significant investments.

The ambition of this work is therefore twofold. On the one hand, it seeks to increase knowledge of Thailand's coastal problems and the adaptive solutions they have produced. On the other hand, it seeks to provide a more comprehensive intellectual and useful resource for communities, scholars, practitioners, and politicians negotiating comparable thresholds in other places. In addition to providing guidance for neighbouring nations, Thailand's experiences can also help shape solutions in a variety of locations, including the Caribbean, the Mediterranean, and the Pacific, where coastal resilience is becoming a more significant concern of our time.

11.2 Shared Shores, Shared Futures: Interconnected Challenges in a Warming World

The climate crisis does not recognise national borders, and nowhere is this more evident than along the world's coastlines. Hence, there are no imaginary political and geographical boundaries when climate emergencies are on the table. The same pattern emerges with startling regularity everywhere from the Andaman Sea shores to the West African deltas, from the brittle Pacific atolls to the expansive urban coastlines of Latin America: populations congregate where land meets the sea, economies rely heavily on coastal resources, and the very systems that support life are becoming increasingly threatened by global warming. Thus, Thailand's experience is a chapter in a global saga characterized by interconnected futures and shared vulnerabilities, and it is a part of a much bigger narrative.

Despite geographical differences, the drivers of coastal vulnerability remain remarkably consistent across regions. As noted throughout the book, as well as in many case study examples in the developing world, rapid urbanisation remains to be one of the most significant drivers of these changes. Cities like Jakarta, Manila, and Dhaka are growing rapidly in South and Southeast Asia, frequently without proper land-use planning, infrastructure, or regulatory control. Similar patterns may be seen in West African cities of Lagos and Abidjan, where uncontrolled coastal urbanization has resulted in the spread of informal settlements over flood-prone areas, increasing vulnerability to sea level rise and storm surges. Similar pressures are seen in Latin American cities like Rio de Janeiro and Lima, where industrial growth, port infrastructure, and tourism-driven coastal development are pushing into environmentally vulnerable areas. Natural buffers like mangroves, wetlands, and dunes that originally absorbed the effects of storms and tides have been eroded in all these instances because of the competition for economic growth at the expense of ecological stability.

Another common theme that unites these areas is ecosystem deterioration. The destruction of mangrove forests in Nigeria's Niger Delta, the bleaching of Pacific coral reefs, and the deterioration of salt marshes in the Gulf of Mexico are all reflections of Thailand's problems with mangrove loss, coral bleaching, and deltaic erosion. In addition to providing habitat, these ecosystems serve as the first line of defence against the effects of climate change. Their deterioration threatens fisheries and livelihoods, eliminates natural protection, and speeds up erosion. Similar to how deforestation and shrimp farming have exacerbated coastal vulnerability in Southeast Asia, the degradation of mangroves for aquaculture and urban expansion has done the same in West Africa. Coastal wetlands in Latin America have been divided by urban development and industrial agriculture, increasing the risk of flooding and reducing biodiversity. The trend is widespread and systemic, i.e., ecological harm exacerbates climate risks by acting as a trigger for vulnerability as well as a symptom of it.

Economic dependency on coastal zones further deepens the challenge. Because they are home to important ports, popular tourist destinations, and industrial complexes, coastal regions provide a disproportionate amount of the world's GDP. A contradiction is created by this dependency: the very activities that propel growth also make people more vulnerable to climatic threats. This is exemplified by global supply chains, which mainly depend on coastal infrastructure. A cyclone in Bangladesh or a hurricane in the Gulf of Thailand affects more than just local economies; it ripples through trade networks, impacting energy markets in Europe, food security in West Asia, and manufacturing in East Asia. Additionally, migration trends highlight how coastal hazards are interconnected. Pressure on neighbouring nations increases when sea level rise uproots inhabitants in Pacific Island states or the deltaic plains of Bangladesh, leading to intricate social, economic, and political repercussions. A local adaption problem soon expands to a regional or perhaps global one.

The field of geopolitics is not exempt from this interdependence. Adaptation to the coast has evolved into a platform for collaboration and rivalry. On the one hand, international agreements and regional alliances are emerging to pool resources, exchange expertise, and develop collective ability. This attitude of cooperation is demonstrated by the Pacific Islands Forum's support for climate justice, ASEAN's coastal management initiatives, and cross-border flood early warning systems in West Africa. However, competition for climate finances, adaptation funding, and technology resources exposes significant disparities. More importantly, poorer and more vulnerable regions/areas find it difficult to draw in investment, while wealthier countries/regions and well-connected cities frequently receive most of the funding. In addition to slowing down global development, this imbalance widens already-existing geopolitical gaps and transforms adaptation from a place of common survival to one of negotiation and power dynamics.

The necessity for coordinated forecasting, cross-border early warning systems, and shared data is one of the most pressing demands arising from this complicated terrain. The dangers of climate change do not end at customs inspections. A changing monsoon pattern impacts fisheries from the South China Sea to the Coral Triangle, while a storm building over the Bay of Bengal poses a hazard to Bangladesh, India, and Myanmar. However, our collective capacity to respond is constrained by data silos, institutional fragmentation, and uneven monitoring. Interoperable technologies that can monitor and forecast hazards at various scales and governance structures that facilitate information exchange and coordinated action are necessary for creating resilient coasts. Without such collaboration, adaption initiatives run the risk of continuing to be reactive and piecemeal, failing to keep up with the rate and scope of climate change.

Lastly, the lesson is clear: coastal resilience cannot be treated as a series of isolated national projects. It needs to be integrated into international trade policy, migration planning, development objectives, and climate initiatives as part of a collective global agenda. One coastline's destiny is inextricably linked to another's. The same climate processes that threaten Kiribati's life also affect the Mekong and Niger deltas, Thailand's mangroves are a part of the same ecological web as Mozambique's wetlands, and Brazil's urbanization is shaped by the same market forces that drive coastal development in Vietnam. Understanding this interdependence is essential for taking meaningful action and is not just a cerebral exercise. The same tides bind the shores we live on, despite their differences. Our ability to behave as though this truth counts will determine the future of coastal societies and, consequently, the future of humanity.

11.2.1 Unequal Shores: Coastal Vulnerabilities Across the Global South and Small Island States

It is important to note that not every shore faces the same future, even though the processes influencing coastal development are similar. Despite being global in scope, the effects of climate change show themselves differently depending on location, growth paths, and sociopolitical circumstances. The gap between the vulnerable landscapes of Small Island Developing States (SIDS) and the coastal areas of mainland nations in the Global South is the clearest example of this imbalance. Although both face existential dangers, there are significant differences in the scope, severity, and consequences of their vulnerabilities, which influence their ability to adapt and their paths to resilience.

Density, dynamism, and development pressure characterize coastal zones of the Global South's mainland regions, which range from the riverine estuaries of West Africa to the enormous littoral cities of Latin America and the deltas of South and Southeast Asia. Cities like Dhaka, Lagos, Ho Chi Minh City, and Lima are situated at the nexus of fast urbanization, socioeconomic disparity, and environmental deterioration, and these regions are frequently home to tens of millions of people. They have multiple layers of vulnerability. In addition to threatening communities, sea level rise also poses a threat to vital infrastructure including ports, highways, and industrial areas. These dangers are exacerbated by subsidence brought on by uncontrolled development and groundwater extraction, which speeds up land loss and increases vulnerability to flooding. Many mainland regions also face intensified storm activity, shifting monsoon patterns, and saline intrusion, which undermine agriculture and freshwater supplies.

Even if there are serious threats, there are also chances for widespread adaptation because of the size and economic importance of these coastal areas. Bangladesh and Vietnam are two nations that have started experimenting with integrated coastal zone management, which combines ecosystem-based solutions with manmade protections. My earlier work in the Southern regions of the Hainan Island in China also reflects on similar development patterns as well as comparable coastal management opportunities. Relocation tactics, delta zoning, and urban planning reforms are being discussed and, in certain situations, put into practice. Mainland regions frequently have better institutional capability, a larger tax base, and greater political leverage to mobilize adaptation financing, notwithstanding governance issues and resource limitations. Even if implementation is still patchy, regional cooperation frameworks, like the African Union's Blue Economy Strategy or ASEAN's coastal efforts, nevertheless provide opportunities for coordinated action and shared learning.

For SIDS, the picture is starkly different. I have in mind to write a book relevant to the context of SIDS and their climate vulnerabilities. Their challenges are already at a new different level. Vulnerability in this context is existential rather than merely a question of exposure. If present sea-level rise trends continue, countries like Tuvalu, Kiribati, and the Maldives might literally lose their land in a matter of decades. SIDS frequently have little economic diversity to support extensive adaptation infrastructure, and they lack the geographic cushion to withdraw inland, unlike mainland states. Due to their frequent reliance on climate-sensitive industries like tourism, fishing, and agriculture, they are particularly vulnerable to both economic shocks and physical effects. Furthermore, without substantial international help, it is challenging to attract investment because to the high cost of adaptation in relation to their GDP. The administrative and technical difficulties of carrying out adaptation initiatives in isolated island contexts can be prohibitive, even in cases when funding is available.

An additional significant distinction is seen in the socio-political aspects of vulnerability. In addition to being an environmental concern, climate change also raises issues of sovereignty, identity, and cultural continuity for many SIDS. In my recent discussions with country leaders (i.e., presidents, prime ministers, and ministers) of several Pacific nations, the topic of land loss, infrastructural loss, and cultural identity loss seem to be of their major concern. Countries and regions like Palau, Kiribati, Marshal Islands, Cook Islands, Vanuatu, Tuvalu, Niu, Papua New Guinea (PNG), and many others, share the same challenges. One leader from the Cook Islands presented it well by saying we are all part of the same family, in the same ocean, with the same challenge. More frequently than before, deep moral and legal questions are brought up by the possibility of climate-related displacement: how can a nation-state maintain its sovereignty when its area becomes uninhabitable? What happens to customs and civilizations that are closely associated with certain landscapes when those regions go under the ocean? Despite all their difficulties, mainland states hardly ever address these issues with the same ferocity or urgency.

At the same time, SIDS have become powerful voices in global climate diplomacy. By means of organizations such as the Alliance of Small Island States (AOSIS), they have influenced the discourse surrounding climate justice by highlighting the unequal burdens they experience in spite of their negligible contribution to global greenhouse gas emissions. Through their efforts, international climate negotiations have placed adaptation and loss-and-damage finance at the forefront. This diplomatic leverage demonstrates that vulnerability is not the same as passivity, especially when combined with creative strategies for cultural resilience and community-based adaptation.

The contrast between island nations and mainland coastal regions highlights a fundamental reality: whereas the factors contributing to climate risk may be similar, the actual experience of that risk is radically different. Therefore, adaptation tactics need to be adjusted to socioeconomic realities, cultural settings, political limits, and physical location. Managing scale and complexity is a challenge for mainland governments, which includes coordinating local efforts with regional and national frameworks, integrating various sectors into coherent adaptation strategies, and striking a balance between ecological restoration and urban growth. Survival is the top concern for SIDS, which includes obtaining financial and technical assistance, protecting their cultural identity and sovereignty, and, in certain situations, making plans for relocation or managed retreat. Countries like Tuvalu are already grappling with this reality on an immediate and existential level.

The climate crisis is remapping the world's coasts in uneven ways, creating new fault lines of risk and resilience. However, these distinctions do not have to separate. If anything, they emphasize the need for solidarity and tailored assistance, acknowledging that a fair and successful global adaptation strategy needs to take into consideration the distinct vulnerabilities and capabilities of every coastal community. Whether they are isolated islands in the Pacific or megacities on deltas, they are all connected by a common fate moulded by the rising oceans. A collective response is both necessary and possible in that shared destiny.

11.3 Translating Thresholds into Practice: Toward a Global Framework for Coastal Adaptation

The complexity and scale of the climate crisis mean that stories, however compelling, are not enough. They need to be transformed into plans, frameworks, and initiatives that can influence governmental choices, direct urban development, and enable local communities to adjust according to their own conditions. Despite being highly contextual, Thailand's coastal experience provides a useful set of lessons that can be applied elsewhere, a sort of adaptable "DNA" that can guide climatic responses.

These lessons demonstrate how concepts based on place-based knowledge, participatory governance, and ecological thinking can be reinterpreted and adapted to fit the many circumstances of coastal societies around the world. They are not about copying, but rather adaptation.

One of the most significant strands of this adaptive DNA is participatory governance. The effectiveness of bottom-up strategies that give local voices and knowledge systems priority is demonstrated by Thailand's coastal adaptation initiatives, especially in smaller fishing communities and peri-urban populations. Adaptation efforts have frequently been successful because they were co-designed with the most impacted parties, not because they were mandated from above. For example, community organizations' participation in mangrove restoration projects has guaranteed both ecological success and local stewardship, creating a sense of ownership that top-down interventions frequently fall short of. This idea is broadly applicable: from Bangladesh's flood-prone deltas to Mozambique's mangrove-lined coasts, incorporating local communities in decision-making is not only a matter of inclusivity; it is a requirement for long-lasting, context-sensitive adaptation.

Equally central to Thailand's experience is the emphasis on ecosystem restoration as a foundation of resilience. A move away from the obsolete idea of "hard engineering" solutions alone is demonstrated by the deliberate rehabilitation of mangrove forests, the development of living shorelines, and the incorporation of green infrastructure (GI) into coastal communities. Although they may be required in some situations, levees and seawalls frequently shift risk rather than remove it. Nature-based Solutions (NbS), on the other hand, provide co-benefits: they buffer against storms and erosion, sequester carbon, support fisheries, and enhance biodiversity. These methods are applicable everywhere. For instance, wetland restoration is being included into urban planning as part of larger climate adaptation measures in Latin America, while mangrove replanting is being investigated in West Africa as a more affordable option to costly constructed barriers. The lesson is clear: resilience cannot be built on concrete alone; it must be rooted in ecological health.

An additional essential element of a framework for transferable adaptation is the use of traditional and Indigenous knowledge systems. In order to manage resources and adapt to climatic variability, coastal communities in Thailand, especially those in the south, have traditionally relied on local environmental cues such as tide patterns, seasonal fish movements, and mangrove regeneration cycles. These knowledge systems enhance adaptive ability and enhance decision-making when paired with scientific facts. In other contexts, this hybrid method has great potential. To improve catastrophe preparedness, Pacific Island countries, for example, are already combining satellite data with Indigenous meteorology and navigation knowledge. In a similar vein, community-led monitoring programs in Latin America are integrating contemporary risk assessment instruments with traditional ecological practices. These illustrations demonstrate how adaptation based on cultural continuity is frequently more robust, acceptable, and long-lasting.

The question of scale is another crucial dimension. Many of Thailand's achievements started out as small-scale initiatives, such as a flood-resilient village or a mangrove cooperative, but with careful integration, their ideas can influence larger policy frameworks. Deliberate procedures are needed for scaling, such as integrating successful pilot projects into national adaptation strategies, connecting regional initiatives to global climate funds, and establishing channels for information sharing. Additionally, it calls for governance frameworks that can handle complexity—polycentric systems that provide local autonomy while coordinating activities across scales and jurisdictions. For transboundary issues like river deltas or shared coastal habitats, where isolated efforts would never be sufficient, this strategy is especially important.

Technology, too, has a powerful role to play, but it must serve rather than dictate adaptation strategies. The way we perceive and address coastal hazards is changing because of developments in artificial intelligence (AI), remote sensing, and predictive modelling. High-resolution satellite imagery allows for more precise tracking of erosion and land-use change, while AI-driven flood predictions, for example, can offer early warnings that save lives and safeguard livelihoods. By improving community engagement through crowdsourcing data collecting and participatory mapping, digital platforms can also democratize adaptation. However, technology is not a panacea. It needs to be rooted in regional reality, influenced by social context, and directed by local priorities. When treated as a complement rather than a substitute for local knowledge, technology can amplify resilience rather than undermine it.

Finally, any global framework for coastal adaptation must be grounded in principles of social justice and equity. Indigenous populations, women, coastal fishing communities, and informal settlements are among the groups most disproportionately impacted by climate change. Thailand's experience demonstrates that adaptation works best when it proactively tackles these disparities by allocating resources, increasing involvement, and guaranteeing that the advantages are distributed. This is a practical requirement as well as a moral one. Adaptive capacity is strengthened by inclusion and weakened by exclusion.

It is clear from the lessons learned from Thailand's tropical thresholds that adaptation is neither a one-off endeavour nor a general model. It is a dynamic, ever-evolving process that combines social inclusion with ecological restoration, combines innovation and tradition, and scales up small triumphs to systemic transformation. These ideas serve as the cornerstone of an international framework that can direct action in coastal regions from Lima to Lautoka, and from Dakar to Dhaka, as the climate crisis worsens. To implement them, we now require institutional cooperation, political will, and a common commitment.

11.3.1 Coastal Resilience Checklists: Six Strategic Pathways for a Changing World

The accelerating pace of climate change and its cascading impacts on coastal systems demand not only analysis but also practical guidance. Although highly specific, the lessons learned from Thailand's tropical thresholds may be condensed into six strategic paths that together serve as the foundation for a paradigm for coastal resilience around the world. These routes are flexible instruments that serve as a compass for communities, planners, and governments navigating unpredictable futures rather than rigid formulas. They exhibit a philosophy of adaptation that is firmly grounded in the ecological and social realities of location, collaborative rather than technical, and iterative rather than linear.

The first pathway, "ecosystem-centred adaptation", recognises that the most powerful defences against climate impacts are often those provided by nature itself. Wetlands, coral reefs, mangroves, and dune systems constitute living infrastructure, not merely decorative landscapes. Mangrove restoration initiatives in Thai areas like Krabi and Samut Songkhram have improved fishing productivity, buffered storm surges, and decreased erosion, proving that ecological restoration may support local livelihoods while promoting resilience. This strategy is equally applicable elsewhere: in Senegal and Mozambique, mangrove rehabilitation is serving as the cornerstone of coastal protection plans, while in the Mekong Delta, wetlands are being incorporated into flood management systems. Through ecosystem-centered adaptation, conservation is reframed as a crucial component of climatic infrastructure that needs to be integrated into development policy, catastrophe risk reduction, and spatial planning.

The second pathway, "participatory and inclusive governance", is about who makes decisions, how they are made, and whose voices shape the outcomes. When imposed from above and disconnected from the lived realities of those most impacted, climate adaptation fails. This is amply demonstrated by Thailand's experiences with coastal management. Participatory planning procedures, which involve discussions with youth networks, women's organizations, and Indigenous communities, have produced more context-specific solutions in fishing villages along the Andaman coast, like co-managed mangrove forests and locally created early warning systems. Adaptation must be developed with communities, not for them. This is a global lesson. This is especially crucial in areas with disjointed governance structures or where past injustices have prevented marginalized groups from participating in decision-making. In addition to increasing legitimacy and local support, inclusive governance incorporates a variety of agendas and knowledge systems into the planning process, producing more creative and just results.

The third pathway, "knowledge integration and co-design", calls for a deliberate blending of traditional ecological wisdom with scientific research and technological tools. In Thailand, more accurate adaption solutions have been developed by combining satellite surveillance and climate modelling with local knowledge of tide patterns, seasonal fluctuations, and mangrove dynamics. This hybrid strategy boosts adoption, raises relevance, and deepens understanding. It also illustrates a more general reality: neither science nor conventional wisdom can effectively address the climate catastrophe on their own. In Latin America, community-led mapping is being combined with geospatial analysis to pinpoint susceptible areas, while in the Pacific Islands, Indigenous forecasting techniques are being combined with meteorological data to improve disaster preparedness. Beyond simply exchanging knowledge, co-design involves working together to develop solutions that are both technically sound and culturally relevant.

A fourth pathway involves "flexible and adaptive urban planning". Static infrastructure models, strict zoning laws, and traditional masterplans are no longer appropriate for a future with a changing environment. Urban coastal areas of Thailand, especially Bangkok and Pattaya, are starting to test out adaptive frameworks that include flexible land-use regulations, green infrastructure, and dynamic risk assessments. These methods enable cities to adapt to ecological feedback, socioeconomic changes, and new climatic data. Similar ideas are being investigated in the floating infrastructure of Rotterdam and the flood-adaptive planning of Ho Chi Minh City. The basic premise is that since adaptation is a dynamic goal, planning must be proactive, responsive, and iterative rather than final and deterministic.

The fifth pathway, "transboundary collaboration and regional frameworks", addresses the reality that coastal systems and climate impacts do not stop at national borders. Countries are connected to common biological and socioeconomic networks by rivers, ocean currents, fisheries, and weather systems. The importance of coordinated action is demonstrated by Thailand's involvement in ASEAN-led coastal programs and cooperative disaster preparedness drills with neighbouring nations. In other areas, early warning systems in the Bay of Bengal are becoming more regionally linked, and the Coral Triangle Initiative brings six countries together to manage common marine ecosystems. In order to build resilience, a cooperative architecture that cuts beyond geopolitical boundaries is needed, including shared data, standardized standards, collaborative planning, and methods for coordinated reaction.

Finally, "financing resilience and scaling innovation" is essential to transforming good ideas into impactful action. Due to financial limitations and disjointed finance structures, many adaption solutions are still only able to be implemented at the pilot level. Green bonds, climate resilience funds, and public–private partnerships are examples of innovative methods that can help unlock the resources required for widespread implementation. Lessons can be learned from Thailand's experience using blended finance approaches to fund community-based mangrove restoration. Kenya's coastal insurance programs and Fiji's migration initiatives are investigating similar strategies. In order for small-scale innovations, which are frequently the most successful and culturally anchored, to scale and have an impact on policy, financing systems must also give priority to local projects.

These six pathways do not exist in isolation; they reinforce one another. In the same way that integrated knowledge strengthens flexible planning, inclusive governance aids in ecosystem restoration. Technology innovation enhances participatory procedures, and transboundary cooperation increases the impact of finance initiatives. When combined, they create a comprehensive structure that is resilient, equitable, and flexible. This checklist of six comprehensive pathways provides a useful road map for navigating an uncertain time, whether it is used in the urbanized shorelines of Latin America, the mangrove deltas of Southeast Asia, or the coral-fringed Pacific coasts. One choice, one shoreline, and one town at a time, it is an appeal to shift from disjointed, reactive adaptation to coordinated, proactive resilience.

11.4 Epilogue: Beyond Thresholds and Toward a Planetary Coastal Ethic

We are no longer waiting for the future to arrive; it is already lapping at our shores. Sea level rise, the swelling of storms, land erosion, and the gradual infiltration of saltwater into formerly rich soils are not merely theoretical forecasts. For millions of people worldwide, these are their everyday realities. Once thought to represent the stable boundaries of our civilization, coastlines are now the most dynamic frontlines of global change. The future of human societies and ecosystems is being written in these shifting margins, and it will be determined by the decisions we make—or do not make—today rather than by the laws of nature.

This book began in Thailand's tropical thresholds, but its lessons belong to the world. These principles are universal. It has proven that adaptation is not a policy paper to be stored away or a technical checklist. It is a significant cultural, political, and ethical endeavour that calls for creativity, restraint, and determination. Coastal areas are more than just places to protect people from the sea; they are places where justice, survival, and environmental/planetary stewardship come together. The outcome of the climate problem will be decided here. Humanity needs to change the way it thinks and behaves in this situation.

A new ethic is required, one that values interdependence above the delusion of control. Coastlines must be seen by this ethic as links rather than divisions between land and water, between human and non-human systems, and between local action and global effects. It must shift from reactive protection to proactive change, repurposing coastal areas as test beds for novel ways of coexisting, governing, and living. It must be based on solidarity above all else. The ability to adapt is still incredibly unequal, but the rising flood won't discriminate between wealthy and impoverished countries. It is not an act of charity but rather one of survival to bridge that gap through cooperative governance, fair funding, and shared information.

The path forward demands courage. Destroying ingrained development patterns that prioritize immediate gain over long-term resilience will be necessary. It will be necessary to integrate technology into more comprehensive social and ecological policies rather than embracing the deceptive comfort of technological remedies. And in a time when waves transcend borders more readily than regulations, it will necessitate reevaluating sovereignty and accountability.

In the end, we reflect on this very important note that the motivation behind this book is not just to document a national story, but to contribute to a global conversation. My next attempts on this topic of research will focus on the Pacific, Small Island Developing States (SIDS), coastal management cases in Australia, as well as success and failures of climate-resilient strategies in practice. Since the climate crisis is a global issue, our adaptation—as well as mitigation—strategies must also be. Thailand's tropical thresholds, as this book's title entails, are a part of a much bigger landscape of change, where knowledge flows across coastlines, deltas and archipelagos share methods, and one place's lessons can become another's lifeline. This book aims to provide not only insight but also optimism by relating Thailand's experience to the difficulties and goals of other coastal regions. It also provides a roadmap for navigating an uncertain future with common sense, ingenuity, and determination.

Let this be our closing wager: resilience is not a finish line, it is a practice, and more importantly, it is a serious practice; daily, local, and inflexibly hopeful. Thailand's coasts/coastlines have demonstrated that change isn't the thunder of massive projects but rather the slow accumulation of smart decisions, e.g., a mayor who pays attention, a youth network tracking tides, a plan modified, a mangrove replanted. The map alters if we extend that ethic: coasts become covenants between people and place, little towns become labs, and deltas become classrooms. Communities that reject fictitious trade-offs and establish authority where they are will co-create the future rather than having it dictated by policy or conferred by technology. Walk away from these pages prepared to lead with care, to teach with a clear message, to conduct research with purpose, to have impact with intent, to learn across boundaries, to prototype in public, and to advocate for the possibilities. The tide is shifting. Together, with bravery, accuracy, and a leaning toward renewal, we should do the same. It is time for transformation, a bold shift toward climate-resilience and reverse some of our unsustainable trends of development.

MIX
Papier aus verantwortungsvollen Quellen
Paper from responsible sources
FSC® C105338

If you have any concerns about our products,
you can contact us on
ProductSafety@springernature.com

In case Publisher is established outside the EU,
the EU authorized representative is:
Springer Nature Customer Service Center GmbH
Europaplatz 3, 69115 Heidelberg, Germany

Printed by Libri Plureos GmbH
in Hamburg, Germany